OZARK ALMANAC

OZARK ALMANAC

A. E. Lucas

Illustrated by Sherri Christian

INDEPENDENCE PRESS

Copyright © 1986
Independence Press, P.O. Box HH
Independence, Missouri 64055

Library of Congress Cataloging-in-Publication Data

Lucas, A. E.
　Ozark almanac.

　1. Natural history—Ozark Mountains—Addresses, essays, lectures.　2. Natural history—Missouri—Addresses,
essays, lectures.　3. Seasons—Ozark Mountains—Addresses—essays, lectures.　4. Seasons—Missouri—Addresses, essays, lectures.　I. Title.
QH104.5.09L83　1986　　508.778'8　　86-3075
ISBN　0-8309-0443-3

Printed in the United States of America

Contents

INTRODUCTION

Although I grew up in the country and loved it, I spent much of my adult life in the city—too much, actually, for a person of my tastes. I could not forget the crisp scent of pine needles, the hovering of dragonflies over a still pond, the evensong of whippoorwills.

At odd times, the voice of my part Indian great-grandmother seemed to ask within me if I intended to pass all my years inside a glass and concrete box.

Henry Thoreau wrote, "What you seek in vain for, half your life, one day you come full upon, all the family at dinner."

That is what I found here in the Ozarks, although not quite the whole of the family as yet. I have never watched a bear feeding, or an otter or a bobcat. But coyotes and bats, deer and woodpeckers, opossums and tree frogs, puff adders, rabbits, turtles, foxes, velvet ants, and many other of the great family of life have visited the woods clearing where my house stands.

Visited? In a real sense, I am the visitor. They belong here. But I am beginning at last to feel at home in this place, too—almost like one of the family.

A. E. Lucas

JANUARY

1

READING MYSTERY TRACKS
IN THE SNOW

Last night the coyotes were howling again up on the ridge.

Newcomers to the Ozarks often do not believe such a near-legendary beast can be prowling about these friendly hills. I was a doubter myself, when I first moved here. At night when I heard their quavering howls from afar, I would turn over in my warm bed thinking with sleepy complacency, That must be a pack of coon hounds. Ah, the Old South!

Now, when I hear the hunting calls of the pack, I mutter, "Coyotes again. Ah, the Old West!" And it's not as easy to fall asleep again. This change of heart occurred because, one winter evening, it snowed. Later that night I was roused from sleep with the confused echo of a noisy din in my ears—squalling or barking, not too far away.

Making the rounds of my windows I peered out suspiciously at dense cloaking cedars and tree trunks black against new fallen snow. Dark branches were white furred. The snow-draped woodpile looked like an ungainly igloo. Ice crystals gleamed on the tips of cedar needles. A sickle moon rode high above the trees. All was silent and still. Then clear as a bell from the wooded ridge above my house I heard, "Yup-yup-hup-hup-HaouooooOOOOO!" Unmistakable! It was a coyote, and nearby. The long drawn howl came a second time, plaintive but also menacing. It was answered by a burst of wild yelping from the distance.

A coyote pack at large—with snow on the ground. Tomorrow, I decided, I would have a look at their tracks.

The morning dawned cold and clear. Before sunrise I was out-of-doors to fetch in wood for the stove. This aim fled my mind instantly when I saw the giant footprints that circled the house and led across the slant of the hill. A very large animal had made those prints, surely nothing of the doggy kin. Loose, fluffy snow had tumbled into the tracks, but here and there one was clear enough to show it had been made by a cloven hoof. Not a horse, then. Too big for swine. Maybe a cow or sheep or deer. I headed inside to phone my near neighbor.

"Janet, is your cow loose?" I asked.

"Lady?" she groaned. "Not again! I'll have to. . . . Thanks." The phone clicked in my ear as she hung up, not giving me time to say it might be a false alarm. So I built up the fire, had coffee, and went about my chores. The sun was well up when, on a trip to empty ashes around the gooseberry bushes, I heard the frantic bellow of a cow and the excited keening of a hunting pack.

The coyotes! Clearly they had found the strayed cow and were harrying her, dancing about the helpless animal in the classic ring, closing in for the kill.

Instantly I was on the trail, following the hoof marks across the flank of the hill. I remembered to set down the ashpan when the tracks joined those of another animal of about the same size. The heifer—fruit of an earlier escape of Lady's—must be loose too. Yes, the two animals had trotted along side by side, sometimes crowding each other to leave superimposed footprints, only to drift apart for small unnecessary detours and back again.

The heifer. Surely the coyotes would try to drive off the mother so they could get at the young one. As if in answer to that thought, the cow's bellowing broke out again, now almost continuous—and imperative. The staccato yelps of the pack grew louder too.

In some far field a tractor labored—an ironic modern counterpoint to the primordial tragedy being enacted in the woods. I ran on, often slipping in the loose snow. The trail I was following turned into a gravel road where tire marks, partly effaced by drifted snow and hoofprints, showed that a truck had passed during the night.

The noise of battle seemed to be coming from just beyond the next bend in the road, perhaps a quarter mile away.

Blood on the snow. I had been running with my head up, looking toward the road bend. When I glanced down it was to see crimson droplets sprinkled along one set of hoof marks, while here and there larger gouts of glistening red had fallen. And the bellowing had ceased; perhaps the wounded cow was saving her breath for running. The canine chorus was very faint now. The tractor roared briefly, then stopped. The silence was ominous. Puffing and sliding, I ran on and finally came around that last bend—to find a scene for which I was not prepared.

A truck was hanging nosedown over the side of a steep, snow-filled gully beside the road, just past a culvert over a creek. Two men were fixing a chain from the truck axle to a tractor. A woman in a windbreaker jacket watched, seemingly near tears.

No cow, sheep, deer, or other hoofed creature was to be seen. Also, no coyotes. But at that instant a pair of large dogs burst from a sumac thicket to dance joyfully about the busy scene. They were familiar dogs: Heidi the grizzled shepherd and a young part-Husky male. They belonged to Janet and Matt, as did the truck in the ditch. The tractor was Benny's; he had the farm down the road.

"What happened?" I asked warily.

Benny grinned. Matt looked up from the chain to say bluntly, "What do you think of a woman who deliberately drives into a ditch?"

Indignantly Janet denied the charge. The truck had slipped. She had been steering at only fives miles an hour while Matt, on foot, had been driving the cow on ahead. Suddenly the dogs, barking wildly, had come racing down the road from the farm. The cow had spooked and turned back, right in front of the truck. To avoid hitting the animal, Janet had swerved. The truck spun sideways, its front wheels sliding over the gully's edge.

Back again at my own house, I recalled the second animal that had trotted companionably beside Lady, and I phoned Owen. Yes, he had already known that that long-legged cow had strayed over to visit his bull; Matt had just collected her. The cow had been badly slashed on the front udders, probably when she leaped her first barbed wire fence to escape. What? No, he had moved the bull yesterday from the hill pasture to a field right behind his house. He had it in view right now. Chuckling over Lady's bad luck, he hung up.

So the second animal was still a mystery. I went back to examine the tracks. This time I followed them in the opposite direction, toward the woods. Here the two had gone single file for a way, then side by side again. But by a big oak, the pair parted company. One set of prints led toward a dense grove of cedar, where a family of wild turkeys sometimes roosts. The other tracks detoured the oak toward the west, the direction of Owen's farm. I chose the western fork to follow.

Then, looking at the trail of a solitary animal for a change, I made a discovery. Comet tails.

Probably anyone who tracks game knows this directional arrow. When a running animal lowers a foot toward the ground, the foot first touches the snow lightly, digging at the start a narrow, then a deeper and wider furrow which ends in the deep footprint which bears the animal's full weight for an instant. The telltale mark left is in the shape of a comet with a tapered tail. This indicates which way an animal is going. Simple! I was following a trail that led into the woods, and I knew it for a fact.

The woods became thicker. Tree trunks were interwoven with wild grape and multiflora rose thickets where monstrous bare canes arched overhead, then raked down at passersby. Underfoot the tangle of dead canes was almost impenetrable. Then I saw bloodstains again, a large gush near my foot. The hoof marks there were scuffled oddly as the cow backed out of the thicket and turned around.

Enlightenment came. There had been not two animals after all, only one. Lady had entered the woods and found the way too difficult. She had then gone back toward the road in search of an open field route to Owens. In the woods, thorns had torn open the clotted wounds from the barbed wire fence. All the footprints had been made by one animal, coming and going.

Heading homeward for a late breakfast, I took a shortcut over the ridge. There I found my big bonus of the day, the trail of a huge coyote.

The animal had loped down the mountainside, his strides long, his pawprints large. There he had leaped over a wild rosebush five feet across, then skidded down a steep bank, plowing up snow.

At one spot he had sat on his haunches behind a tree, watching some unknown prey. Finally he had passed through a sassafras grove, where a single turkey feather lay in the snow. Had he been able to reach up and snatch a low-roosting bird? Surely these were the marks of the scout coyote, he of the doleful wail in the night under a sickle moon.

By then the snow was melting fast under a bright sun, or I would have followed him to his lair.

2

QUAIL MAKE GOOD PETS TOO

The bobwhite is one of my favorite birds. I like to hear its cheerful whistling from the hillside—this invisible woodland companion of my many rambles.

In other ways, too, this bird is popular. Because of its delicate flavor, it is a table favorite among people who relish game birds. Its excellent camouflage, plus traits like lying well to a dog and making a lightning take-off when flushed, endear it to sportsmen.

Moreover, the bobwhite can make a tame, affectionate pet. The usual way of acquiring one is to discover an unhatched egg in a newly deserted nest. The hen quail lays a clutch of twelve to twenty white eggs, most of which hatch at the same time. As soon as the feathers of the young ones dry, the parents lead them away from the then dangerous nest area.

My brothers and I once found such a nest in which, among shattered eggshells, four intact eggs remained. Were they duds, or only late hatchers? Fortunately the sun had kept them warm. We hurried home with our finds, wrapped them in flannel, and placed them in a shoebox on the back of the kitchen woodstove. To our delight three of the eggs hatched.

The wee bits of dark fluff easily learned to eat chick-starter and drank water on their own. But they were just too lively and curious about the world to stay in the box, so shortly they were exiled to the small-chick yard beside the barn.

There was no keeping these escape artists fenced in, of course. They found new ways to get out every day. However, they shunned the woods, and endeared themselves to my dad by policing the gardens efficiently— eating slugs, ants, beetles, caterpillars, crickets, grasshoppers, spiders, and snails without damaging the plants as chickens do.

Friendlier than chickens, the young quails enjoyed riding around on any available shoulder or extended arm, like pet falcons, or ran behind us cheeping merrily. The contrast of their warm interest in humankind with the chickens' businesslike indifference, and the wary distrust of the half dozen ringneck pheasants we once hatched, was very marked. The pheasants stayed wild birds at heart, but the little quails became tamer than long domesticated barnyard species like ducks and geese. I can only attribute this to the quail's friendly and sociable nature, as my family made no real effort to tame these foundlings.

Those who have tried—usually with a single orphaned quail—report no trouble in gaining a delightful pet who likes to live in the house, roost on a dresser at night, share juice and buttered toast at breakfast, and cuddle on someone's lap while the family watches TV.

The bobwhite (*Colinus virginianus*) is a plump, stubby bird with ruddy mottled feathers, hard to see against dead leaves. It lacks the stately head plume of the western species like Gambel's, Mountain, Scaled, and Harlequin quails. The bobwhite male is slightly smaller than the female, and the sexes are further distinguished in that the male has a white throat and eye bar, while those areas on the female are buff.

Although it is a ground nesting bird, the bobwhite takes pains to make a cozy, well-hidden nest. The female first excavates a hollow and lines it with dried grasses. Then the surrounding vegetation is drawn overhead to make a dome that is bound together with grass and other material. As she sits closely on the nest, facing the sole entrance, the mother bird is very hard to see. Both parents care for the young, and either will pull the old broken-wing trick to lure intruders away from the nest.

By autumn the male bird no longer sounds his familiar call from the edge of his domain. The little family that all summer has boldly ventured into fields, along roads, and even into barnyards, now becomes secretive, wary, and very quiet. Families join to become large coveys that are the target of hunters.

Now, if we are lucky, we will hear the quail's other call—the slurred two-note whistle that is the signal for the covey to regather after it has scattered before some danger.

One peculiarity of the bobwhite covey is its habit of roosting on the ground at night in a close circle with all the birds facing outward. If disturbed, each takes off in a different direction, to the bafflement of the attacker.

The quail is nicely insulated against cold and wet by its unique double feathering. Except for those of the wings and tail, its feathers are double, each having a long smooth outer feather and a short fluffy one underneath, both growing from the same shaft. In severe winter storms, though, even the hardy quails appreciate an uprooted tree with the dried leaves still on it, under which they find not only shelter but often a meal of dormant insects or egg bundles. The landowner who wishes to preserve wildlife will let such trees lie over winter.

Since the quail's normal diet of insects, weed and wild grass seed is so beneficial to humans, some farmers protect nesting areas on their land, limit the numbers of quail shot, and even feed the birds. Quail will come to a feeder area if there is cover nearby, such as a dense thicket, but prefer a field at the edge of a woods.

Verne E. Davison, an expert in birds' feeding habits, suggests lespedeza as a good choice for a food patch, for when competing species clean out the more desirable seeds, the lespedeza will be left for the quail to survive on. A food patch of only a quarter acre will support a covey of twelve of these useful and interesting birds.

For the last couple of years I have been checking abandoned quail nests, but haven't found any unhatched eggs. That's all right, though. The quail families that share my acres and use the water hole below my house are pretty tame as is.

3

A WATER HOLE IN WINTER

A thin edging of ice rims the tiny coves and sandbars of the old water hole. Here in this microcosm of a lake, four stepping stones make a chain of islands. A hummock of sedge thrusting above water provides a perch for a lively chickadee looking for food.

A silken brown cocoon fastened to a tall sedge stem draws the bird's attention. But one side of the tiny bundle has been torn open, perhaps by the white-footed deer mouse that lives in an alder thicket upstream, and the occupant of the cocoon is gone.

Cold wind roughens the surface of the water, pushing

fallen oak leaves toward a partly submerged log at one end of the pool. The chickadee departs in a nervous flutter of wings.

I put down my binoculars, start to turn from the window toward my waiting typewriter. But a flash of red catches my eye, and I lift the glasses again. A pair of cardinals have arrived, the male leading. He alights in a lopsided young mulberry tree on the bank, and looks around cautiously. Seeing no threat, he escorts his dun-coated, pink-billed lady to dine on the rose hips of the multiflora bushes, the cane tips of which drag in the water. Beyond, branches of dark green cedar are decked with pale blue, aromatic berries. The male cardinal flies there, takes a berry in his beak, and offers it to his mate. She accepts, placidly.

A small, humpy black shape darts across the snowy bank, stops dead, then moves forward again. Alarmed, the cardinals disappear into dense cedar foliage.

But the black shape is only a tiny squirrel. He digs furiously. The ground is not frozen beneath the light coating of snow, so he finds a hickory nut he buried last autumn and pushes it into a cheek pouch. Three times more he digs before finding a nut for the other cheek. Then he scampers back to his tree to dine in the warmth of his den.

His home tree is a post oak, partially uprooted as a sapling. Now it leans out over the stony creek-bed, curving down like a rainbow at the far side. This is a tree form my dad used to call an Indian bridge. There was such a bowed tree along a woods trail near my childhood home, where I used to dodge under and out of the shadow of it fast for fear an Indian brave would drop down on my back.

The water hole is deserted now, so I go to the kitchen for coffee.

My house, ideally located for wildlife watching, stands on a rocky ledge overlooking Flat Rock Creek. During winter thaws and heavy rains, a white-water torrent churns down over the wide stone terraces that give the place its name. But most often the creek bed is dry, since the land here—as in much of the Ozarks—is underlaid by limestone riddled with crannies, if not a cavern. Thus, small as it is, the pool is the only permanent body of water in a twelve-acre woods and is frequented by the entire furred and feathered population of the hollow.

More visitors arrive at the pool. No binoculars are needed to study the new arrivals, the turkey flock of the hollow, fifteen strong. The speckled, slim-bodied youngsters run eagerly to the pool edge but are driven back by the hens. Last summer the chicks were pampered, allowed to quench their thirst at once: now they are nearly full grown and must learn manners.

The big tom who leads the flock does not drink yet either. He parades the bank with a lordly strut, feet lifted high. His chest is thrust out, his wattles engorged. His head swivels from side to side in continuous alert.

Why write about the water hole in gray winter? Well, when leaves are out, intervening bushes and low hanging tree limbs completely obscure the view from my window. In summer I do sit by the pool to enjoy the hovering dragonflies, the orange-headed turtle submerged in water to her eyes, the stiltlike waterbugs whose feet seem to dimple the surface of the water and whose ominous shadows overhead cause schools of tiny tadpoles to flee in panic for the cover of the log.

There is also a shiny black snake about four feet long that patrols the banks, but seems not to swim. No big animals are to be seen, though. And not even blue-jays—the boldest of birds—will land to drink or bathe when I am nearby. I sigh, and wait for winter.

It was in late winter that a young doe came sedately in midmorning to part the water briskly with her feet, pushing aside dead leaves before lowering her head to drink. There was snow on the ground when a spotted skunk, its belly fur so long it nearly swept the ground, ambled out on the partly submerged log (the wild bees' drinking place in warmer weather) to bask in the pale sunlight for an idle hour. And last, a silver fox—that rare and secret animal—came to frisk beside the quiet pool.

A small, long-legged animal with tiny black feet, the fox moved gracefully, picking its way over rough terrain. His coat was tinted luminous pink by the sunset, his black tipped brush carried jauntily. He lapped his fill, sniffed the air again, then ran partway up the old Indian bridge tree, led on by the enticing smell of squirrel. Before he reached the top of the arch, though, he seemed to remember pressing business elsewhere. Backing up, he dropped lightly to the ground and vanished into the woods.

It must be admitted that a very important life form is missing from my water hole. There are no fish—not for long, anyway. During winter thaws, when the creek runs full from bank to bank, fish do wash down from upstream. The last one to trust the seeming haven of the quiet pool was a bass nearly a foot long. Misty, the big female raccoon, showed up and caught him in broad daylight the same day. Her kits were not with her, but I knew they would have a fine dinner when Misty, her catch in her mouth, reached the den.

To be honest, sometimes I yearn for a big lake such as local farmers have—lakes leaping with fish, where a whole herd of deer may mingle with drinking cattle and hundreds of migrating wildfowl set down in passing.

Still, the old water hole has quality. How many folk can behold from their living room windows the glory of a silver fox, its tail semaphoring mischief, fur pink-washed by the glow of the evening sky, come in to water at sundown?

FEBRUARY

4

THE INDIANS CALLED HIM ORCHEK

A cold wind rustles the dry leaves still clinging to an oak where a flock of purple-hued grackles are huddled. A few snowflakes dance in the air. A hungry rabbit gnaws bark from a staghorn sumac. Last night, traces of aurora borealis flickered in the sky.

Orchek has never seen the aurora borealis, and is oblivious to the bleakness of the winter landscape. Down in his snug subterranean burrow he lies wrapped in the long sleep of hibernation. His heartbeat is slowed; breathing is nearly imperceptible. His body temperature is stable at about fifty degrees, the same temperature as the surrounding earth below the frost line. From such profound slumber, biologists say, it takes a hibernating animal several hours to rouse.

Will Orchek awaken as expected on February 2 (his species' national day) to see his shadow? This year I intend to loiter near the entrance of his burrow, just to see.

I do not like to call this dignified wild friend of mine a groundhog—that hodgepodge of a name. If he lairs underground, so do rabbits, foxes, moles and a host of other animals who do not have "ground" affixed to their names. And he is a true marmot, not even distantly related to the hog. He does not wallow; he has furry feet and not hoofs; and his eating habits are finicky (he is strictly vegetarian except for an occasional grasshopper). His other title, woodchuck, sounds better. It is thought to be a garbling of the animal's Indian name, which was Orchek (or Otcheck, in some dialects).

He was Orchek to me from the start. That was a warm day in late spring when, glancing out my window, I was startled see a pudgy reddish animal standing on his hind legs, breaking down tall weeds so he could munch the tender top leaves. His head was broad, small-eared, large-eyed. His face was blunt, with two large front incisors glinting in the sun. His body was about the size of a beaver's and of the same shape, but the tail was short and brushy.

Something about that erect animal silhouette touched my imagination. I had just been reading about early exploration of the West, and it was of thousands of prairie dogs—standing sentinel duty at the entrances of their vast underground cities—that I was irresistibly reminded. Now why couldn't this visitor of mine be the last of the free-living prairie dogs, miraculously surviving here in the Ozark hills? But no, it was only a common groundhog *Marmota monax*, and I felt disappointed.

Then the blood of my Indian (Lenape) forebearer asserted itself in me, and I smiled. *Orchek* had come to visit: my lodge was honored.

This placid browsing animal, also known to the Indians simply as Digger, was often mentioned in old campfire tales. If he lacked the color and dashing style of a coyote, trickster deity of the Plains Indians, Orchek the Woodchuck Spirit was a respected character in the forest community, helpful and benign. In old folktales a hunter meeting Orchek in a forest glade might seek the spirit animal's advice or tips on weather changes. Lost children might be taken in and sheltered by Orchek in his roomy underground home on a cold night.

Interestingly, the woodchuck's roles of weather prophet and shelterer have passed over into the white man's culture. Each February 2 people all over the nation listen to news broadcasts to learn if Punxsu-tawney Phil, the famous Pennsylvania groundhog, will see his shadow, thus predicting a late spring. Since the animal would come out of hibernation in midwinter only after an unseasonal warm spell, which Ma Nature usually makes up for with severe weather later in the winter, this makes sense.

As for providing shelter, the woodchuck is a fast, per-sistent digger of extensive tunnels, where most any wild animal may take refuge when pursued. Also rabbits, foxes, skunks, weasels, and opossums will make their homes in deserted woodchuck dens.

Orcheck makes a main burrow with several side tun-nels, some leading to craftily concealed outside en-trances, others blind branches. A clean animal, he may use one of the blind branches as his latrine. The central den is a large chamber lined with grasses and dried leaves.

Excavated dirt, piled outside the front door, often makes a mound of some size, on which he will sun him-self after feeding. Frequently at sunset he may be seen standing there erect, as if on sentinel duty, gazing alertly but contentedly over his front yard.

A touch of mystery clings to the woodchuck. He is the Digger. But is he also the Whistler? Several farm wives had told me that when there is a woodchuck about, they hear a kind of low-pitched short whistle, ama-teurish, "like an eight-year-old boy just learning." I heard this sound for the first time after Orchek began his visits.

Biologists say that the woodchuck does make a shrill whistle when alarmed, but what I heard didn't sound excited or fearful (to my ears, anyway), only shy and tentative. In fact, just like an eight-year-old boy practicing.

A harmless and interesting animal, the woodchuck has one fault—he is attracted to gardens, especially in early spring when food is scarce. Thus I watched to see if Orchek would raid my patch. My plants are the kind that come up on their own every year and need little care—asparagus, rhubarb, parsley, spearmint, and catnip—with a sprinkling of self-seeding cherry tomatoes and a tall row of Jerusalem artichokes. Orchek ignored all these delicacies, seeming to prefer weeds.

His favorite food is a milky-juiced plant that grows about six feet tall from a basal rosette of jagged leaves like a dandelion's. Curious, I checked with a plant guide and found it to be wild lettuce, perfectly edible and sought for by wild wood experts like Euell Gibbons.

So far, Orchek and I have a truce. He lets my garden alone, and I forbear pulling out his wild lettuce when I occasionally tidy up the yard.

Will he awaken on schedule this year, to see his shadow?

For his sake I hope he doesn't, because he'll be lean and hungry. And there's not a darned thing around now for him to eat but the bark of trees.

5

MAPLE SYRUP, INDIAN STYLE

Mention maple syrup and people think of the north woods, especially Vermont. Yet in earlier times this delightful "woodcraft" was practiced widely in the South, from the Great Smokies and the Ozarks down to Georgia and west to Texas.

Not just the sugar maple but all species of the maple family, box elder, birch, and even sycamore trees were tapped. Their sweet sap was converted into syrup and small brown cakes of flavorful sugar. Later, tree tapping in the South became a lost art, displaced by the sorghum industry.

Nevertheless, making maple syrup can still be practiced here as a hobby, an outdoor recreation activity for the whole family, or a pleasant way to combat cabin fever in that drab time of the year when it seems that spring will never come.

The American Indians, who pioneered tapping trees, revered the maple as a special gift of the Creator. Gathering sap was an exciting time that culminated in a great spring festival in which quantities of maple and birch beer were quaffed by the menfolk between athletic dances in a kind of forest carnival. Some of the syrup was poured on clean snow to make maple "taffy"—a treat for children and adults alike—and some of the fermenting "beer" was set aside by the squaws to become maple vinegar, relished on vegetables and broiled meats.

Syrup was sometimes carried on the trail in emptied duck eggs—the early disposable container—to be mixed with hominy for a high energy snack. But the more usual transportable form was the familiar brown cakes of sugar, a lively trade item in colonial times.

The Indian's fondness for maple sugar was not, at first, shared by English settlers. One even wrote that the product "lacks the pleasing, delicate flavor of cane sugar." Maybe this reaction was simply sour grapes, because Indians had discovered the goodies first, for maple products seem to have a wide appeal otherwise. Bees and other insects have been observed drinking sap, whenever a maple's bark has been newly injured. Deer will lick at the oozing sap for its sweetness. Birds are often seen clinging to the spiles in tapped maples, drinking their fill.

Often the snow was so deeply drifted in the Pennsylvania woods that my dad had to go to his sugarbush with sledge and team. Our horses dipped their muzzles into sap buckets every chance they got. Our dog Tippy not only drank sap but also loved maple taffy. My brother was always slipping her a chunk, on which she would chew for hours. Tippy's distress, when she could not get her stuck-together jaws apart for proper barking when visitors appeared, was pitiful to see.

The Indians' way of collecting sap, which they demonstrated to the early colonists, was to cut a deep V slash in the bark of the tree, just as the Indians of Mexico and the Amazon do today in tapping rubber trees. They drove a pointed stick in at the lowest point of the V; where the sap concentrated and dripped into a bark trough.

Henry Schoolcraft, visiting Indians on the shores of Lake Superior in 1823, saw the sap being stored in large vats of hide stretched over wooden frames. He observed that the whole tribe happily joined in the activity of "sugaring."

Legend has it that in the golden days when the world was young, Indians could draw syrup right from the tree without labor. But some dunce of a culture hero, fearing people would become lazy, climbed up and poured water down inside the trees. Since boiling down the sap, the tedious part, was squaws' work, this fellow ranks as an early antifeminist. But the women were not buffaloed for long. Finding a half-completed dugout canoe, they poured the watery sap into it and left it overnight. In the morning they lifted the thick cake of ice off the top to find golden syrup filling the trough beneath.

Ever since then certain tribes have used the shortcut method of freezing the water out of the sap when making sugar. Of course the freezing gives a syrup that is still too thin, so it must be finished off in large earthen pots into which hot stones are placed. This is slow, but it does work. Such stones hold their heat for hours, and the syrup is not likely to burn, as may happen so easily when a metal pot is being used over open flames.

My kid sister, who lives in Michigan, has used the freezing technique at times. It does markedly shorten the boiling time. There is some loss of flavor, but this can be overcome by blending in a syrup made the long-boiling way. And maple is a strong flavor, too, so this additive may not be necessary for many palates. The maple syrup bought at grocery stores usually contains only 3 percent maple syrup, the rest is corn syrup.

Primitive as it sounds, the Indian method of sugaring was very productive, where maples or birch were plentiful. In 1870, E. L. Sturtevant wrote that the Winnebagoes and Chippewas sold the Northwest Fur Company as much as 15,000 pounds of sugar a year. He also remarked on the carnival air of sugaring time among the tribes.

If you'd like to try your hand at tapping, and don't have all the stainless steel equipment of the modern Vermont sugarbush, consider trying the old-fashioned way. No need to slash your trees' bark as the Indians did, though. Shallow holes can be bored with a brace and bit. For spiles, cut elderberry wands into ten-inch lengths, push out the soft core with a wire, and insert the spiles into the holes in the tree trunk. If the weather is right, sap will soon fill your buckets.

When is the right time to tap trees in Missouri? February is customary, but any time from a January thaw until the leaves come out, as long as the days are warm and sunny, about forty degrees, following crisp nights with temperatures in the middle twenties, will give some sap flow.

Half the fun of sugaring time is getting out in the woods with a lively crew, breathing the fresh air with a tang of woodsmoke, exercising heartily, and keeping warm over the steaming kettles and pans. The other half, certainly, is in that good maple flavor. Months later, as you pour the bubbling amber syrup over a pancake stack and watch it melt the pat of butter, you can savor anew the experience of tree-tapping time with each ambrosial mouthful.

6

THE CARDINAL: A BIRD FOR ALL SEASONS

A welcome visitor at bird feeders this time of year is the cardinal. His bright red coat enlivens the most drab of winter landscapes.

Unlike the bluejay, the cardinal has good manners and will allow smaller birds to feed unmolested near him. Also, his voice is better than the strident jay's as he calls "What cheer, cheer, cheer, cheer, CHEER" from the top of a tree. As a variation the cardinal will sing his other song, "Purty, purty, purty, purty, CHEER!"

Originally a tropical bird, as its colorful plumage suggests, this species (*Richmondena cardinalis*) has been extending its range steadily northward. Once known as the Kentucky Cardinal and the Virginia Redbird, it is now at home even in the Great Lakes area. Where cardinals go, they stay year round, proving themselves a hardy and efficient breed.

The cardinal's endearing habit of singing even in gloomy weather was noted by Audubon.

How pleasing it is, when, by a clouded sky, the woods are rendered so dark, that if it were not for an occasional glimpse of clearer light falling between the trees, you might imagine night at hand, while you are yet far distant from your home—how pleasing to have your ear suddenly saluted by the well-known notes of this favorite bird, assuring you of peace around, and of the full hour that still remains for you to pursue your walk in security.

A similar experience befell me once in a Michigan forest, and that ringing "What cheer, cheer, cheer . . ." really did stir me like a voice from home.

Surprisingly the female cardinal sings too, although her singing season does not last as long as the male's. In late February or early March, if the weather is mild, the cardinals will be looking at nesting sites and singing of territory and true love. With an early start and good luck, a pair may rear four broods a year. In fine teamwork the female incubates the eggs while the male brings food for her, and later for the young ones too. Maturing quickly, these young will be ready to leave the nest in eight or ten days. Fairly soon the male will take charge of them so that the female may start a new brood.

Like many of the grosbeak family, the cardinal is primarily a seed eater, although he indulges in select insects and the cotton worm. Sunflower seeds, cracked corn, sesame, and peanuts are sure to attract him to a feeder. Chopped apple, in lieu of summer's berries, is another cardinal favorite. In the wild he will dine on fox grape, blackberries, pokeberries, chokeberries, the fruit of dogwood, black haw, and sarvis, plus filberts, pecan, and hickory meats, but the seeds of assorted plants and grasses form the bulk of his food.

This diet of the cardinal once gave me a bad hour. My

cat Smith sometimes brings me the gift of a mouse or mole—alive, often unhurt, as she seems to think I will want to play with it. (Smith can be as gentle-mouthed as a golden retriever of my acquaintance.) One day, to my dismay, Smith was waiting on the doorstep with a baby cardinal in her mouth.

Perhaps it had fallen from the nest but it might well have been taken during that first perilous flying lesson, as it was of good size.

I could see no wound on the bird, but it stayed motionless in my palm, resting on its side, feet held in flat against the abdomen, eyes bright, and its little heart racing. When I raised or lowered my hand, it made no effort to spread its wings to fly off.

In the toolshed I keep an old bird cage, in which I once raised a baby robin. But the robin readily ate raw hamburger and bits of fat trimmed from pork chops, while I had a hunch that at this stage a young cardinal would be fed on predigested (regurgitated) seeds, something out of my power to arrange. Regretfully I laid the limp feathered bundle on the flat roof of an empty wren house on a metal post, and went indoors with the cat.

From the window I watched anxiously. For a long time the bird lay still. Then, slowly, it gathered its feet under its body and crept to the very edge of the birdhouse roof. Again it was motionless, for nearly half an hour. At last, though, it uttered a forlorn little peep.

Soon the peeps became louder and shriller.

It was not long before the parent birds found their missing offspring. The interrupted flying lesson was resumed. The first attempted flight was a fiasco, with the baby merely fluttering to the ground right beside the post. The second try was better; it coasted a couple yards to land on a thin lilac branch about a foot from the ground. There it clung timidly, wings outspread for balance.

My heart was lighter as the parents led the little fellow off into the woods.

I like to think this rescued fledgling made it through that rigorous first year of life, and that it is one of the adult redbirds eating at my feeder now. If so, by the end of February another cheery voice will be tuning up for the courting season.

MARCH

7

TRAILS IN THE WILDERNESS

There is a magic about woods trails.

For example—after struggling through unfamiliar terrain, over mossy logs and fallen branches, all at once your feet find easier going. You have hit on a well-beaten path that seemingly came from nowhere to carry you on your way. No longer distracted by a need to fight through thickets and bramble, you begin to enjoy the scenery.

Tree trunks soar upward like pillars in some primeval temple. Delicate ferns, splashes of wildflower color, some oddly shaped rocks arranged with almost Japanese artistry, please your eyes. And up ahead, in a sun-filled glade carpeted with wild mint, a close encounter of the bird kind may be waiting for you.

If you stay with this little path it may take you just where you wanted to go—a spring of clear water to cool your thirst, a blackberry patch ready for plundering, or simply the edge of the woods and a recognizable landmark if you were...well, not lost, but maybe a bit "turned around."

Such mysterious trails in the wilds gained a special interest for me when, as a kid tagging along behind my dad through a forested area, we struck what seemed a much used thoroughfare. The path was worn inches deeper than the surrounding ground.

It happened that as the path led past a mighty beech tree, it ran directly under a low gnarled limb that caught Dad about chest-high. Casually he placed a hand on the limb and vaulted over. Too small to try that, I crawled beneath, bumping my head.

"Of all the dumb. . ." I muttered. "Why didn't they take a saw and cut off that limb when they were making a path?"

They? Suddenly I realized that whoever made that path didn't need to cut off branches, because it could amble underneath without even ducking its head. The trail had been made by the woods dwellers themselves. It was an "animal highway."

True, such paths may have been used, extended, and linked up with other trails by bands of primitive people. However, if there is a low pass over a mountain range, a waterhole in the desert, or a sluggish spot in the current where a river can be forded, animals will have found it first and blazed a trail. One suspects that those "trackless wilds" of early travel books never existed. If the explorers thought so, that is a sure sign they were fighting the terrain, trying to navigate solely by compass.

Any mention of woods trails brings to mind the classic single-file path winding through the forest through which Daniel Boone and his Indian friends may have jogged. That's not the whole picture, though. Whether or not there were larger total populations of big animals in the old days, the herds could at least migrate freely during droughts or severe winters in search of food, instead of being confined in isolated sectors as is the case today. The track made by herds of deer, antelope, etc., in their seasonal wanderings was doubtlessly as wide and well-marked as the swathe left by the caribou of the Far North or the old-world reindeer herds today. Control of one such game migration trail across Kentucky is said to have sparked a bloody Indian war, as did choice ambush sites along the regular routes of western buffalo herds.

Indian trade routes and game trails were often utilized by white settlers moving through the Appalachians and beyond. Widened and smoothed for wagon traffic, later paved as highways or graded for the railroad, with fords bridged and ravines filled, many of these routes still serve us today.

In the West, historical travel routes like the Santa Fe Trail, Navajo, Bannock, and Chisholm trails are celebrated in legend and song. Interestingly, the beautiful Lolo Trail, the Indians' favorite over the Great Divide, is still negotiable only by foot or on horseback (or possibly by four-wheel drive), as are parts of the old Bannock Trail of Montana that are not used for cattle drives.

It must be said that modern hiking pathways like the Pacific Crest, Appalachian and Ozark trails are not of animal origin, except perhaps for short stretches. No animal ever needed to make a trail along the crest of a mountain range. And bands of primitive men would keep strictly to high ground only for military purposes—espionage and raids.

At any rate, it was near the Bannock Trail just north of Yellowstone that an animal path once saved my life.

On an elk hunt, I had gotten separated from the others and found myself on a mountaintop looking down into the very valley where our camp was. The route by which we had come, around the mountain, was six rugged miles, but here I could see the campfire far below.

Through field glasses I picked out old Shorty, cutting up venison for the evening stew. The coffee pot was already on, and I could almost smell the aroma. Naturally, I started right down the mountainside.

Distrusting the jumbled boulders of the scree to my left, where an unlucky footstep might dislodge a rockslide, I began to descend a wooded section of slope. It was steeper than the scree, but those pines looked mighty sheltering. Alas, pine needles are also slippery. Within minutes I was tobogganing on the seat of my pants, faster and faster. Then came the abrupt shock as my hurtling body wrapped itself around a tree trunk. Feeling for broken ribs, I caught my breath and looked around. Below me (I had come to a stop on a knob) was a drop-off of ninety feet. Above me, the ascent looked nearly perpendicular, impossible to climb.

Then my eye fell on a narrow yellow line that slanted across the rocky scree. A wild goat path! Scooting sideways like a crab, I reached its safety. The path led to a salt-lick on the valley floor, so I strolled down that mountain easily, in record time.

Maybe the moral of this story is, some trails you can spot only when you are close to ground level, if not reclining. When seated on a log recently, I was able to observe the tiny pathways and foraging runways made by a deer mouse around its home, which had been invisible when I was standing up. The paths had been designed to take advantage of every overhang of rock, grass tuft, or weed, to hide the comings and goings of the small reaper from the sharp eyes of hawk or owl.

Foraging was even safer when winter snows roofed over the mouse's runways, creating tunnels where its trips were well concealed. But the snowbank had melted now, while the wind had torn away much of the sheltering vegetation, exposing these secret byways.

Even the paths of mice and voles are mammoth, easy to spot, compared with the courses of the carpenter ants whose hive I all but tripped over in the woods one summer day. How like the aerial view of a human city it was—those gray weathered spires (the remains of a tree stump) that housed thousands of individuals, many busy coming and going over an arterial network that radiated in all directions.

Traffic was two-way on these crowded single-lane highways, so that the threatened collisions of what looked like tiny black gas trucks was almost continuous. Usually they managed to squeeze past each other but sometimes, on colliding, both ants would be swung around so that each was headed back the way it had come—still at top speed.

And on a sandy stretch crossed by one major ant highway a hereditary enemy, the ferocious larval ant-lion, was busily digging a lethal pit.

Which brought home to me another peculiar fact about woods paths. Whether for cougars, boa constrictors, scalp hunters, or elephant poachers, trails are grand places for an ambush.

Here in the Ozarks, of course, one can stride down a woods trail without fear of such perils. But if there is fur trapping in your region, wear sturdy boots, and keep an eye out for spots of disturbed earth, a glint of metal, or telltale stakes in the trapping season.

Also, watch out for those low-hanging tree limbs.

8

SPRINGTIME AND SASSAFRAS TEA

My memories of springtime long ago on Dad's hill farm are tied up with the taste and smell of sassafras tea. Trips to the woods to dig roots, my breath steaming in the crisp cold air. The family dog Tippy that always ran on ahead of us, barking just for the fun of it. A song sparrow's fluting. The delicate green of fern fiddleheads, the yellow of dogtooth violets. All these come thronging into my mind.

Afterward, there were cups of hot, fragrant sassafras tea sweetened with honey, sipped in the old farm kitchen with its wood stove and blue Dutch plates on the wall. The tea warmed our insides no matter how chill the weather outside.

Wild plant expert Oliver Medsger once wrote:

Half the pleasure of using green foods is gathering the material—following along hedges, fence rows, or by the brookside, or even into the deep woods. A good observer sees much more than the plants he is hunting, and the time spent in the open air gives one an appetite that makes any food taste good.

For me this was truer of sassafras expeditions than gathering greens when I was a child. The trees were big, easily located, and besides we knew where they were from previous years—while patches of small things like mushrooms, Indian turnips, and dandelion greens were in a different place each year. We had to have our eyes on the ground and our wits about us, not to miss them. Perhaps that is why the annual outing for sassafras roots is so indelibly fixed in my mind, while memories of those other outings are vague.

The best root-digging spot on Dad's farm was a narrow spur of mountain that ran down between two fields nearly to the creek. Three generations of farmers had plowed up stones in those fields, and had tossed the stones onto that spur, until hardly an inch of earth showed. But the dense grove of sassafras that spread along the spur didn't seem to mind; the slender boles slid up through the heaped rocks and grew thicker every year. The more we pruned at their roots, the better they grew.

Dad and Big Brother would spade up the roots and chop them into short lengths. We younger kids shook off most of the mud and carried armsful of roots down to the creek for that first washing in clear, cold water that smelled of moss and pine needles. Our work pace was leisurely. We always had time to play rowdy games, and to catch tadpoles in the creek. Or we would just stand still, looking and listening to the woods.

Later in the spring, searching under the sassafras tree for a mitten-shaped leaf the size of our small hands was fun. As its name (*Sassafras variifolium*) indicates, the leaves have varied shapes. One has a large and a small lobe resembling a mitten. Others are oval, or have three lobes. But early in the season, the leaf buds would just be swelling, and we liked to chew on the twig ends and leaf buds (the filé of New Orleans gumbos) as the Indians did. They left a clean, spicy taste in our mouths.

All parts of this tree are aromatic. It was the "Spice Tree" to early colonists, and tons of sassafras roots—our first plant export to the Old World—were shipped to Europe. There it was popular for centuries as a substitute for the more costly cinnamon. (In fact, sassafras is a member of the same family as cinnamon.) For the Spanish, though, this was the "Ague Tree" from which they extracted a medicine to reduce fevers and the swelling of rheumatism. The Indians also brewed up pain relievers and fever cures from root, bark, and twigs, as well as using various parts for flavoring foods, thickening stews, and making sweet drinks.

At any rate, when we got home with our sacks of sassafras root, the first tea of the season was brewed and enjoyed. Then the remaining roots were given a good scrubbing in the sink under a red-handled pump, dried on top the wood stove, cut into shorter lengths, and tied in small bundles. These Dad took with him on market day to be sold to housewives in town.

There were men who went about the countryside buying up roots and herbs for big companies, too. (Sassafras is the flavoring agent of root beer and other soft drinks, candies, chewing gum, toothpastes, and some medicines.) But Dad never had any dealings with these men. The townfolk who bought his roots did so because they wanted to drink the tea as a tonic to repair the ravages of winter to their systems.

We always kept an ample supply of dried roots at home, too. Besides the tea, which all of us liked, we kids also wanted hard candy. (The storebought version of this came in the shape of little root beer barrels.) We made it in the same way as peanut brittle, then cracked it up into bite sizes after it cooled. To make this candy we used an extra strong tea. We kids would fix ourselves homemade root beer with the leftovers. We'd take a half-cup of the strong tea, two tablespoons of honey or maple syrup, and fill the glass with cold spring water. After stirring this well, we'd add a pinch of baking soda to make it bubble. It had twice the flavor of bought root beer.

Old-time country doctors used to prescribe saffrole, the active agent in root bark, for patients with heart conditions and digestive upsets. But in 1926 this drug was dropped from the U.S. Pharmacopoeia, along with hundreds of other herbal remedies.

That was discouraging enough for old-timers who liked their spring tonics, but a couple of years ago worse news came: According to the FDA drinking sassafras tea might tend to cause cancer.

I was deeply shocked. Could it be that those old country doctors, the Indian wise men, and all our grannies were wrong? I would rather believe that the samples used in that study were dug beside a polluted roadway, or that the specimens got contaminated at the lab. Well, perhaps in due time the FDA people will revoke their harsh edict, as has happened with saccharine and nitrates.

In the meantime, springtime without sassafras tea? It just won't be the same. But at least they can't take away those wonderful memories of how it used to be.

9

THE POKEWEED RITUAL

When it's springtime in the southland, thousands of people clutching paper bags and knives, fan out across the countryside in search of pokeweeds.

This highly esteemed wild green is found in clearings, at the edges of woods, and along fence rows. Extra lush clumps await the reapers of nature's bounty around former barnyards. Await? No—*beckon*, for the exotic weed may grow six or more feet tall, and the previous year's stalks of an eye-catching reddish purple will still be standing. One need only walk up and slice off the succulent new sprouts just above the ground at the base of the old stalk.

Sprouts are best harvested when less than six inches high and the tiny leaves are still tightly furled. At this stage they resemble asparagus spears. They don't taste like asparagus, of course. Poke greens have a taste all their own, and a good zesty flavor it is, too.

When I was a kid, though, I didn't care much for any kind of greens. My interest then was in the plant's berries, which make a good dark ink. Because this was much used in colonial times, the pokeweed had a romantic appeal to me.

Naturally, we concocted the ink with much excitement. Whispered code words and potent spells of the Tom Sawyer variety were not for ordinary writing such as school papers. No, our special pokeberry ink was saved for writing notes to the Jesse James gang and Robin Hood's band, warning them of the sheriff's activities. These notes we hid in a hollow tree in the woods. Seemingly those Brothers of the Owlhoot Trail always got our messages, for the notes would be gone from their hiding place when next we looked there, and the local sheriff never did manage to arrest any of our heroes.

It was pokeweed as a green, however, that gained my attention in later years. The ritual of gathering and cooking it was explained by the Indians to the early settlers, who welcomed this addition to their drab diet. Pokeweed seeds were carried to Europe, and the plant is now cultivated in France and North Africa. But Americans can still gather all they need from the wild.

Interestingly, some Indian tribes made an arrow poison from pokeweed too—most likely from the root—but they did not share this bit of know-how with their white neighbors.

Although the thick, fleshy root of the poke (*Phytolacca americana*) is known to be poisonous, the shoots and young leaves are quite safe to eat. Cases of poisoning have occurred only when amateur gatherers, not initiated into the ritual by old timers, have taken part of the root while collecting sprouts.

Also, it is common practice to boil these greens through at least two waters—some cooks throw away as many as five—just as a precaution. There is no need, though, to avoid eating this delicious, mineral rich food just because some other part of the plant is toxic. That would be like rejecting rhubarb pie because the leaf of that plant is toxic. For that matter, the foliages of the potato and tomato are not safe to eat either. In fact, very few plants are edible in all their parts. So pokeweed sprouts are indeed safe as long as one remembers to cut them off about a half-inch above ground level to avoid getting any of the root.

How should pokeweed be cooked? Place the sprouts in a saucepan, cover with water, and boil ten minutes. Drain, cover with fresh water, and boil for another twenty minutes. Dress with salt and butter or bacon drippings. A cream sauce, grated cheese, or even a dash of vinegar may be used on the greens for variety.

Even when the sprouts are a foot tall, they can still be used, according to Euell Gibbons. He suggests peeling off the tough skin, then slicing the white inner core into short lengths. These "logs" may be dipped in egg, rolled in cornmeal, and fried. Or they can be boiled and served with melted butter. Either way, they are delicious. (Removing the skin eliminates the need for parboiling.)

In late summer the stately pokeweed's arched branches dangle long sprays of whitish blossoms, followed by attractive purple berries. Wild birds and chickens can eat their fill of these berries without harm. But childen should be warned against eating them, as the seeds contain a toxic agent.

Many old-timers, though, regarded pokeweed berries

as a tonic, when taken in small amounts. Grandpa always kept a spray of dried pokeberries on a kitchen shelf. When he was feeling a bit unsettled he would eat a couple of these dried berries, much to Grandma's alarm. He lived to a ripe old age, but whether this was due to the pokeberry tonic or to his general feistiness is unclear.

Over the years, my own interest in the berries has waned. It's been a long time since I wrote to Billy the Kid with crimson berry juice. Remembering Grandpa's tonic, however, last year I dried some berries and ate one. (Ugh.) Since then, my trust is placed in the tonic effect of the sprouts. Those once despised greens taste pretty good to me now.

APRIL

10

WHEN WHIPPOORWILLS CALL

By the second week of April, old timers say, the whippoorwills will be returning to the Ozark Mountains.

These birds are not quite the harbingers of spring. Down by the creek the frogs have been piping the overture for weeks, opening the season. But it is the distinctive nightly chorus of whippoorwills for which we all wait.

Persons who grew up in mountains anywhere in the eastern half of the country get nostalgic about the call of the whippoorwill. During hectic years when I was tied to a city desk, my memory often harked back to twilight time at the old farmhouse. My dad used to sit on the porch, contentedly listening. The swelling bird chorus would start up softly on a distant hillside, then resound triumphantly across the purpling valley, to ebb away in faint sweet echoes. Seated lower on the wooden steps, we kids would turn our heads to look up at Dad when some other nightbird called, and he would say quietly, "barn owl" or "nighthawk" or whatever it was.

The fragrance of lilacs and rambler roses from the side of the house, or of new mown hay from the fields, would be wafted in by a south breeze, and the boards under our bare toes would still be warm from late afternoon sunlight. The moon would rise, big and orange against an indigo sky.

It was the best time of day, to a farm child tired from chores and play, stomach filled with a good supper, yawning sleepily as the music of the whippoorwills rose and fell in the dusk.

The whippoorwills' evensong was one memory I yearned to recapture when I moved to a house in the Ozark woods. Things weren't quite the same, though. Oh, I welcomed the first two hours of their chorus. But that first night, as I lay wide awake in my new lodgings well after midnight, waiting for the birds to quiet down (hadn't they ended the chorus as soon as I tumbled into bed in the old days?) I grew impatient, then provoked. Bird calls mellowly echoing from a far hill are one thing, but when hundreds of the shrill-voiced creatures are right outside an open window the noise level can be deafening.

Tossing sleeplessly I felt that, if my well had been put in then, I would have connected the garden hose to spray the trees and drive away the racket makers. Their predawn noise peak was the most strident din I had ever heard.

What was especially irritating was some nonconformist who could not get the song right. "Whip-per-WILL, whip-per-WILL," the massive chorus would chant for five minutes, then cease. In the interlude before the next outbreak this loner would call, "Whirr-WITTa, whirr-WITTa," over and over, as fast as he could.

Once he moved to a nearby tree, and I could catch a

breathless "huh" or cluck before each call. Then his valiant, misguided effort would be drowned out by the orthodox singers, but he never gave up on his version of the song.

I named this loner "Little Dummy." Everyone knows some animal that has not mastered the communication signals of his kind—the cat that can't purr, the dog that doesn't bark at strangers but sings a doleful tenor when a fiddle is played, the hen that sneakily lays her eggs in the bushes with never a squawk. By his perversity Little Dummy raised my particular wrath: if it hadn't been for him, maybe the hypnotic monotony of the others might have lulled me to sleep.

Later, I found out that I did this bird an injustice; the loner was *not* stupid.

Perhaps it was the third night that, exhausted, I dropped off the minute my head hit the pillow. Amazingly, since then I've never had a bit of trouble sleeping. When visiting kinfolk complain that the birds keep them awake all night and ask how I ever get any sleep, I say, "If you lived here, you'd soon get used to it too. Just background noise. It's even restful, because it proves there are no marauders around."

I noted, one summer, that the loner had finally found a lady bird who liked his style; together they raised a family. Now there was Son of Dummy singing staunchly beside his errant father. The two were still sadly outnumbered, though.

The answer to this mystery came to me by chance, as I was leafing through a bird manual. All my life I had considered the whippoorwill just a voice in the dark, invisible. Vaguely I thought of this nocturnal bird as black, like Poe's raven. But no, there was a color photo of it. The bird was invisible for another reason entirely. It was dappled, checked, speckled; brown, cream, buff, gray, and black. Perfectly camouflaged against the dead leaves, it rests silently by day, then becomes active at night when it feeds on insects and carries out its vocal courtship. Nor do whippoorwills engage in telltale nest-building activities. The female lays her two dappled eggs in a slight depression on the forest floor, where they, too, are invisible against dead leaves. Even the chicks, when hatched, blend in perfectly with their surroundings.

Then I recalled that more than once a speckled, nondescript bird, long of wing and short-tailed, had startled me on early morning walks by flying up almost in my face, before vanishing into the trees.

That was the elusive whippoorwill.

Interestingly, the whippoorwill was unidentified by early settlers. They never saw the bird either, and attributed its spooky calls to its cousin, the nighthawk.

But the "poor-will" family has other members, too. The answer to the puzzle of the loner was on the next page of the bird manual. In the South there is a species called chuck-will's-widow, because that is what its song sounds like to some ears. The last two words, which I heard as "Whirr-WITTa," are uttered clearly, but it takes a good ear to catch the initial cluck. Not a dummy, my loner was singing the correct song of his species.

There is now a small colony singing the dissenting refrain in my woods. Last summer when my visiting brother-in-law swore he didn't get a wink all night because of the birds, his young son added, "And did you hear the bunch that couldn't even get the song right?"

"Those were not ordinary whippoorwills," I told them. "They were chuck-will's-widows."

The magic has returned to twilight time for me, now. I like to sit on a lawn chair, listening to the whippoorwills tuning up each evening. The tame 'possum Paula comes out for her portion of catfood. Fireflies flash their tiny lanterns. Up on the ridge Misty the raccoon chirrs softly to her kits as they hunt. A fox barks in the distance. Frogs are still piping down by the creek, but their place will be taken soon by cicadas, those other musicians of the night, adding a skirling rhythm to Ma Nature's orchestra.

An orange moon as big as the ones of my childhood climbs in the sky.

I find the evening serenade very restful now, and soon my eyelids grow heavy.

11

THE DOGWOOD, TREE OF LEGENDS

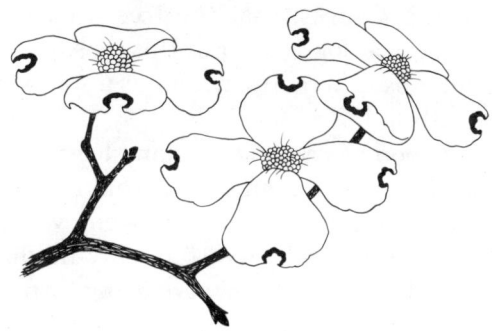

In springtime the flowering dogwood (*Cornus florida*) makes a fine show, decorating the woods with white clouds of blossom when many other trees are still bare.

On a leisurely walk through the woods at this time, my senses are stimulated with signs of new life. There is the rich damp smell of leaf mold underfoot. A song sparrow's warble competes with the clear whistle of the bobwhite. Pale, lemon-colored leaf bracts enfold the swelling hickory buds, and the flamingo pink of a few newly opened oak leaves glow like tropical flowers. But of all the signs of earth's seasonal rebirth, the dogwood in full bloom is the most visible.

This is not just another tree; it is something special.

A hardheaded, otherwise unsentimental Ozark farmer once said to me, "I hate to see a dogwood cut down. Won't cut one myself, if I can help it."

When asked why, he hesitated. "Well, for one thing the birds feed on the berries. Quails and such."

Now it is true that the fruit of the dogwood, and its twigs, are important food for wildlife. But that wasn't the whole story, with this farmer. "Besides," he glanced up at the sky, a revealing gesture, "that little tree *means* something. It's hard for me to put the idea into words, though."

Later I discovered he was referring to a curious legend of the Ozarks, that the tree used to make the Cross had been a dogwood. For this reason the dogwood is sometimes used here in sermons as a symbol of rebirth and everlasting life.

The dogwood has a place in many legends. According to one of my favorites, which was told me when I was a small child, long ago the Devil went walking in the woods. He saw a small tree staunchly holding onto its pure white blossoms, which were being torn at by a stiff breeze.

"You are the smallest tree in the forest," he said, smiling. "But I can make you the biggest and mightiest of all, towering above the oak, so that even the winds of a tempest won't bother you. Just do homage to me."

"Never," answered the dogwood. "Don't you see that each of my flowers has four petals that are arranged in the shape of a cross?"

Furious, the Devil took a sharp bite out of every one of the white petals. And to this day if you look at the dogwood flower, you will see a tiny nibble is missing from each petal. And the edges of this bite are brownish, as if scorched by a great heat.

At one time I wondered if at least one member of the dogwood family had made a pact with the tempter. The Pacific dogwood (*C. nuttallii*) has petals with entire, unmarred edges, and its height may be eighty feet or more. But where this tree grows, overshadowed by giant redwoods, sequoias, and lodgepole pines, it is still the smallest tree in the forest.

Legends aside, the graceful dogwood holds a secure niche in the hearts of people and is often transplanted to lawns and gardens. Both the white and rarer pink-flowered varieties occur in the wild and have not needed much improvement by nurserymen. Purists like to remind us that the real flowers of the dogwood are tiny yellow things in the center, while the showy white "petals" are only modified leaf bracts. But that is true also of the flaming poinsettias, and in no way detracts from the beauty of these flowers.

The dogwood's leaves, which grow in pairs opposite each other along the stems, are oval with prominent curved veins and turn an attractive scarlet in autumn, providing a second colorful show each year.

Slow growing, the tree has a dense, hard, fine-grained wood that is prized for making shuttles for weaving.

The berries are of good size, nearly half an inch in length, and occur in clusters. They, too, turn bright red as they ripen in the fall. These provide food for not only the bobwhite quail but many songbirds, including cardinals, summer tanagers, mockingbirds, and wood thrush.

Interestingly, two northern cousins of this tree are also valuable to wildlife. One is the bunchberry (dwarf cornel) a low-growing plant whose berries were eaten by Indians, birds, and bears. The leaves are an important part of the ruffed grouse's summer diet. The larger, shrubby red-twigged dogwood (cornel) has edible nourishing fruit harvested by Indians, bears, pheasant, and grouse. Moose and elk feed heavily on the twigs and winter buds. The strong, Y-shaped limb crotches were used by Indians for slingshots and cooking racks.

Where I now live there is a slender young flowering

dogwood in full view from my windows. A pair of cardinals nest in a nearby cedar, and the male claims this dogwood tree as his own territory, singing his brisk "Cheer, cheer, cheer" from its topmost branches by day. At night the light of the spring moon touches the white blossoms with unearthly radiance.

Gazing out at it, I quite agree with that Ozark farmer. The dogwood does mean something special that is hard to put into words. I'll never cut one down either.

12

A WELCOME FOR SWALLOWS

The saying goes, "One swallow does not a summer make." But two of them can, if they are a devoted pair of barn swallows that return every year to a mud nest plastered under the eaves of a house.

That's the way a grateful farm family here in Jake's Prairie feels. The wife says, "Since the swallows have been nesting at our place—that's four years now—we can sit on our porch on a warm evening, or have an outdoor barbecue, without a single mosquito bite. Before, we risked being eaten alive by bugs."

The glad news that there were swallows in the area filtered to me through a neighbor who has a feeder and is fairly familiar with Ozark birds.

"It's strange," she said one day, "that my new martin house stands empty save for sparrows, but a lovely purple martin pair have a nest on the side of my friend's house, up under the eaves.

"At first when the martins kept bringing mud pellets to stick against the side of her house, as soon as the nest got heavy it would slide down. The wood is unpainted, but has been treated with raw linseed oil, you see.

"So the man of the house nailed a board under the nest for support, and soon the pair were feeding some voracious youngsters. The birds catch bugs while they are flying in the air. It's something to see."

The martin is a swallow too, but the description of nesting habits made me think that the neighbor's welcome guests were barn swallows. The two birds are rather alike in color. The purple martin, a larger bird, has a shining violet head and shoulders, shading to bluish black on the rest of the body, while the female of the species has a grayish beast and white belly. The barn swallow is a rich blue-black with rufous underparts and a chestnut splash on the breast and forehead; the female has duller colors and a shorter tail.

It is easy to distinguish these two species in flight. The martin's wings appear more triangular and, of course, its tail is not deeply forked like the other's.

Wangling a visit to my farm friends, I found that they already knew they had barn swallows, not purple martins, and were quite proud of them. The birds had raised only one brood the year they built the nest but had produced two broods in each of the years since. That day the birds were feeding heavily above a cornfield, alternately flapping and coasting, with side trips out over a nearby lake.

The nearness of such a body of water is critical to the swallow's choice of a nesting site because of the need for wet mud. But the greater concentration of insects around water is important too.

Songs and folklore are rich in references to the swallow. There are few species more beloved or more useful.

And when people speak of the swallow they mean our familiar barn variety (*Hirundo rustica*) noted for the beauty of its soaring, dipping flight.

Its trademark is the long, deeply forked "swallowtail" that shares its name with an attractive tiger-striped butterfly as well as to a curious fashion in men's formal dress, the swallowtail frock coat of bygone years.

The barn swallow displays itself to best advantage in flight as it gathers insects in midair with its flat short triangular beak. Its wings are long and powerful from constant use, while its weak feet are used only in perching. It is the beak that is used for carrying mud pellets.

The nest is a marvel of patient industry, a cup-shaped structure built of tiny gobs of mud, smoothed and mixed with dried grasses for added strength. The outside appearance is that of a flattened haystack. The inside of the nest is lined with soft feathers. Four to six eggs are usual; the shells are white with red-brown spots.

While the barn swallow is sometimes confused with the purple martin, it is even harder to distinguish from its near relative the cliff swallow. This bird, which bird watcher champion Hal Harrison suggests might better be called the eaves swallow, once built its nests on ledges and crannies of steep cliffs, before the coming of the Europeans. Since then the bird has adapted well to civilization, fastening its mud nests on the sides of buildings under the eaves.

When open barns were commoner in our countryside, a nest on the outside was often that of a cliff swallow, while the barn swallow preferred to build inside on beams and rafters. Audubon reported once counting more than fifty nests in a single barn. Barn swallows are social birds, and where conditions are favorable a colony will spring up, with many nests in rows, only inches apart.

Even in Audubon's time the species chose other sites too. He observed birds nesting "under bridges, or sometimes even in an old well, or in a sinkhole, such as those found in the Kentucky barrens."

When a hidden cave is discovered through a glimpse of a bird flying up out of a hole in the ground, that bird is usually a swallow.

Cliff swallows are also sociable, form colonies with mud nests all-in-a-row, and resemble the barn variety in size and color. How can these close cousins be distinguished? Again, by viewing them in flight. The cliff swallow is square tailed. It also has a noticeable buffy rump patch. Actually, it doesn't matter which of these three species is attracted to a place; all of them are assets. They live on an insect diet and are fast feeders. All three maintain a close, almost symbiotic relationship with people. As human populations increase, too often wildlife dwindles. So far, this has not been the case with some species of the swallow family. Farmers everywhere have cherished such valuable allies in their war with bugs.

The Indians especially appreciated purple martins and put hollow gourds on poles for nesting at their garden plots. In Holland and other European countries, clay pots are fastened sideways to the walls of houses for swallows to use. In America special attention has been paid to the martins that like to live in colonies, and roomy avian "apartment houses" on stilts now grace many lawns. It is not necessary to do anything special for cliff swallows, it seems, as they can find ready nesting sites on almost any building.

For barn swallows, the big red barns of yesteryear were ideal. This species thrived and grew numerous. But now that these open, rambling structures with their timbered haylofts are becoming rare the birds have come on hard times. For this reason, anyone who buys an old farm with weathered buildings should consider leaving some of them standing as a haven for this now diminishing species.

Swallows winter in Mexico or South America, where they are affectionately called *las golondrinas*. In April they return to Capistrano, to Saxonburg in Germany, and to quiet farms in the Ozarks, where I can cajole an invitation to come and sit on the porch to enjoy one of the most graceful sights in nature, the soaring flight of a swallow.

It doesn't take much to welcome a swallow—a ramshackle building, a pond or lake, or sometimes just a board nailed under a sliding mud nest.

MAY

13

THE ART OF RAISING SQUIRRELS

Not long ago my niece wrote me that a new member had been added to her household—Merle the Squirrel. She rescued him from a neighbor's dog, after the baby squirrel fell from the nest. He now lives in a specially equipped birdcage in the living room and is gaining weight on infant formula.

"It reminds me of home," my niece added happily. "Mother was always raising squirrels."

Her mother—my kid sister—indeed had a unique relationship with these bright-eyed, friendly little creatures. The birdcage in Sis's living room often held a baby squirrel that had fallen from a nest, and sometimes even an injured adult that she was nursing back to health before setting it free.

There was Bozo the Clown, affectionate Tykie, shy Pixy, robust Bushy Tail, lovable Weenie the Helpless, and others whose names I have forgotten. The house where Sis lives is in a grove of maple and beech, with a large squirrel population, so when she hears frightened whimpering in the grass she gets a towel at once and goes to the rescue of yet another foundling.

Her love affair with squirrels, which dates back to her own childhood, did not have a pleasant start.

As it happened our big brother, then a teenager, took his .22 rifle and went to the woods to hunt. He coaxed Sis to go along and gave her the job of entering one side of the woods, waving her arms and shouting to get the squirrels' attention, while he sneaked in from a different angle and shot them. The plan went beautifully. Yet as Sis tagged home behind the young hunter with his bag of squirrels, she was very quiet. At supper, for the first time, she had no appetite for squirrel potpie.

Even if this hadn't happened, Sis probably would have been the type to rescue and mother little orphans of the wild anyway.

Of all those orphans, Weenie the Helpless was the most difficult to handle. When found crying weakly in the grass, Weenie looked like a newborn. Toothless, of course, but also nearly hairless, pink as a wiener, his eyes still sealed shut. It was about a month, I think, before his eyes opened. Sis wrapped it in flannel, placed it near a hot water bottle for warmth, and fed it with an eyedropper. Later she used a dolly bottle. He did well on diluted formula and nursed eagerly.

So very young and helpless was Weenie (even more helpless than a human baby that can at least void by itself from its hour of birth), that the squirrel had to be helped with toilet training. After each feeding, gentle stroking of the lower abdomen and bottom with a cotton ball moistened in warm water initiated this vital function, a result the wild mother accomplishes by firmly washing up her litter after feeding them.

Within weeks his fur had grown in beautifully. He soon progressed to soft foods. Cherries, oatmeal with cream, carrots, apples, and pieces of doughnut smeared with peanut butter were his favorites. When his teeth appeared, though, he showed no inclination to use them. A peanut hull he could get through, but even a thin-shelled English walnut baffled him. If we cracked the nut for him, he would nose out the meats and eat it, but that was his limit. He showed no interest in acorns or hickory nuts. Whether his refusal to learn about the natural food of his kind was due to laziness (everyone spoiled him) or merely from lack of an adult role model, as Sis thought, I am not sure. Unfortunately the children of the family stuffed Weenie with sweets and junk food, which didn't help any.

So his teeth grew. And grew. Those strong front incisors of the gnawing family grow out constantly to make up for the steady wearing away of daily use. Finally Weenie could barely get food past them into his mouth. Sis's husband had to clip the edges off those steadily growing teeth with his grape pruning shears from time to time.

As with all their squirrels, Weenie had the run of the house. The cage was merely his sleeping place, secure from teasing by the family's dogs and cat. Eventually he did learn the joys of gnawing, and the old organ in Sis' living room still shows the marks of his chisel teeth. He would also nibble on sticks and twigs placed in his cage, or on any convenient furniture. Always on wood, though. Weenie never once chewed on a hard-shelled nut.

Now my niece writes that Merle the Squirrel, on the contrary, is almost too interested in nuts. Brazil nuts are his favorite, and he can make it past the hard shell in no time. He begs for nuts, far more than he can eat, and hides them in the cedar chips in the bottom of his cage. When out of the cage he robs the nut bowl and even fetches nuts from the yard (given him by the neighborhood kids) to hide under the cushions of chairs and sofas.

That's not such bad news, I pointed out to her. It sounds as if Merle is developing the right instincts to survive in the wild on that day when he's big enough to be released.

As I write this now at my desk, I can glance out the

window and see a sleek, free-living fox squirrel scampering along a tree limb. That's Sue the Singer—one I raised. If I go outside with a peanut and call her, she will come down to a lower limb and accept it from my hand, but she is skillful enough at finding her own food.

Sue has two voices, a loud scolding one for when she's upset, and a soft chirruping croon that she pours into my ear while sitting on my shoulder. (*I* think she's trying to sing, anyway.)

Actually I wasn't the one who found Sue crying under a den tree. Taffy the mother cat carried in the rain-soaked waif by the nape of the neck, put it in the box with her own kittens, washed, and nursed it. After a while I removed the newcomer from the box and hunted out eyedroppers and cotton balls, but not because I feared Taffy would forget herself. The baby squirrel, although its eyes were still closed, was nevertheless bigger and more aggressive than the four-day-old kits and was getting a lion's share of the chow.

As I watch Sue frisking in the treetop now, I remember Weenie the Helpless, the squirrel that was terrified of the outdoors, and could never have been set free anyway because he never learned how to crack a nut.

And I can't take all the credit for Sue's independent ways either. True, from the time she opened her eyes, I took her for a walk in the woods each day. Then she found a last year's hickory nut on the ground and just started nibbling.

14

A HUMMINGBIRD IN THE GARDEN

There are hundreds of hummingbird species in the Western Hemisphere, most of them in the tropics. Of the handful that venture north of the Rio Grande only one, the ruby-throated hummingbird (*Archilochus colubria*) is commonly seen in Missouri. It is a summer visitor, usually arriving in May.

Brilliant in coloring, this tiny bird has iridescent green plumage with white underparts, and adult males have a flaming crimson throat. Attracting such floating jewels to one's garden has high priority with many confirmed bird watchers.

A neighbor of mine, who succeeded, reports a very interesting phenomenon she calls "signaling," in one hummingbird pair that returns to her yard every year. The woman spends much of her time working in a rear room where a large window overlooks the woods from which the hummingbirds emerge daily, during the summer, to visit her feeder.

If the hummers' special hourglass-shaped feeder runs empty, they will go at once to this large window. There they fly repeatedly from the bottom of the glass pane to the top, spiral out and downward, then soar up again until they have caught the attention of the room's occupant.

Interestingly, during the ascending part of this flight pattern the male bird keeps his front toward the window as if showing off his best feature, the vivid throat patch, to anyone inside. This resembles somewhat his behavior during his spectacular courtship "pendulum dance." But of course he could just be looking in the window, too.

At one time I lived in a third-floor city apartment. On snowy winter days, throngs of sparrows would perch on my windowsill and peer inside, chattering noisily while they awaited my table scraps and bread crumbs. (I didn't know much about running a feeder in those days.) But just outside that window was the fire escape landing, their usual feeding spot. Now in the case of the hummers, their feeder hangs from a tree limb in front of the house, while the large window where they make their aerial display is around at the rear—even the door through which someone emerges to refill the feeder is on another side of the house.

Are the hummingbirds signaling?

A higher authority I consulted on this matter did not think so. An explanation he offered was that the male sees his reflection in the glass and makes a hostile display to that "other bird."

One problem with this explanation is that the female shares in the demonstration at the window—and there is no reflection. A sheet of plain glass is not a mirror, it reflects an image only when rays of sunlight strike it at the right angle. But this window, due to heavy shading from trees, is never hit by the sun's rays in the summer. What is seen from outside the window is the inside of the room with its constant occupant, the bird lover who refills the feeder. (I verified this at several times of the day.)

Further, the birds appear at this window only when the feeder is empty. (I witnessed this, too.) It is hard not to think that there is a cause-and-effect relationship here.

At present, my guess is that what is involved is a type of anticipatory behavior in animals. For example, if a farmer oversleeps, and his livestock miss being fed at the usual hour, they may not wait at the usual feeding place. Frequently they will line up along the fence where they can watch the house or see the coming of the truck. Outdoor pets that are not confined will gather near the door. If a pet is inside the house it may come into the bedroom to see what is wrong.

Once chanced upon and rewarded—the sight of the animal may remind the human of the overdue feeding— the pattern will be reinforced with each successful repetition.

Whatever the true explanation of the hummingbird's impromptu aerial ballet at the window, I am grateful to my bird-watcher friend for sharing this experience with me.

If the feathered folk kept a book of records, the hummingbird would hold many of them. It can fly straight up or backward or hover stationary in midair. Hovering occurs during feeding, when the wings are beating nearly sideways instead of up and down, which seems a violation of the laws of flight.

Incredibly, this mite of a bird with a wingspan of less than five inches can achieve a speed of a mile a minute. That's with a tail wind, as one ornithologist pointed out, but its more average speeds of twenty-five to forty-five miles an hour aren't exactly dawdling. Hummingbirds in their regular migration make a flight of hundreds of miles nonstop across the Gulf of Mexico without rest or food, a heroic feat for a being that doesn't weigh much more than a wet bumblebee.

Nobody can win all the titles, though. A hummingbird is not a songster. The humming sound that gives the family its name is made by the rapid whirring of the wings. The male's voice, heard most often during the courtship dance when he swings in pendulum-like arc before the eyes of a watching female, has been compared by one observer with the chattering of an excited mouse.

After the nuptials, the male returns to a bachelor existence. He does not visit the nest the female has built of plant fluff, bits of soft lichen, and spiderweb, saddled on a tree limb. Measuring about an inch across, this little cup is room enough for the usual two white, unmarked eggs the size of navy beans. Incubation and feeding the youngsters are handled by the female.

How can hummingbirds be attracted to a yard?

In my friend's case, the magnet was an old-fashioned honeysuckle vine that grew over the porch. When she saw a pair of hummers visiting the honeysuckle she rushed out and bought them a feeder vessel. It was that simple.

Although the ruby-throated hummingbird does eat small insects, the larger part of the diet is nectar. The long slender bill is admirably suited for siphoning nectar from deep, funnel-shaped flowers. Some favorite flowers are trumpet vine, scarlet runner bean, and morning glory. One or more of these vines running rampant on your back fence will surely attract hummers in the vicinity.

According to Verne E. Davidson, an authority on the feeding habits of wild birds, if you want to landscape for hummingbirds, you might include in your permanent plantings some of the following: mimosa, coralberry, weigela, and the azaleas. With garden flowers, the same ones visited by bees will also draw hummers. Dahlias, gladiolus, tiger lilies, four o'clocks, beebalm, sage, snapdragons, petunias, phlox, and zinnias are among those listed by Davidson as nectar producers.

Once you spot hummingbirds in your garden visiting the flowers, by all means hang up a feeder to supplement their diet. These birds, as their rapid movements indicate, have a very high metabolic rate and may visit the feeder as often as every fifteen minutes. Be sure to hang the feeder near a window or on the patio, so you can enjoy the sight of these colorful visitors to your garden.

The hummingbird is unique in another way also. It's the only bird with a feeder that stays in the closet all winter, and is brought out when summer comes.

15

PONDS AND KIDS GO TOGETHER

Those men of science who study inland waters, limnologists, define a pond as a body of fresh water that is small, quiet, and shallow.

A pond will have a mud bottom where plants may root, with their tops above water. It will be rimmed with plants too. Unlike a lake it has no tides, no waves to disturb its surface. Its temperature is unstable, though, varying around the clock as the air temperature changes. Therefore the amount of dissolved oxygen in the water also varies widely, determining which kinds of fish can live there.

Kids, who are inseparable from ponds, have their own definition.

"If I can wade across and keep my shoulders dry," my

nephew says, conceding quickly that there may be a couple of deep holes for fish to lurk in during the heat of the day, and where a person might get a good ducking, "and there are frogs and little fishes and tadpoles. . . ."

That's the essence: a watery habitat teeming with interesting life forms, but scaled down in size to suit a child or an indolent nature lover like me. A pond is a friendly, welcoming sort of place. It does not have suddenly hostile, threatening moods as do larger lakes or the sea.

For kids, there are fascinating discoveries to be made at a pond's edge. Seasonal changes to note, like the slow sliding of spring into summer. How many kinds of plants and animals live there. How they interact as a community. Who eats what in the food chain.

Red-winged blackbirds, which nest among cattails and rushes, eat the seeds of marsh plants and meadow weeds, also small fruits and insects. The little bass and bluegills darting about in deeper waters have a diet largely made up of larval insects and small crustaceans. Wildfowl eat the fish but also will dine on grass seeds, wild celery, watercress, duck potato, sago pondweed, and yellow water crowfoot.

There are dominance patterns to observe, too.

Cattails, sedges, and bur rushes abound. Clearly they are the dominant plant life of the shoreline and shallow zones, as pondweed is in deeper waters. Several species of duck feed almost totally on the latter. Great submarine jungles of it can form, offering a haven for small fish, snails, and so forth.

But when the tangled masses of pondweed hinder the pond's owner in his boating, swimming, or fishing, it may lead to the pond's being drained and cleaned. This is a catastrophe for many small creatures that dwell there. In time, though, these members of the little aquatic community will reestablish themselves, or their ecological niches will be filled by other species.

Trees are part of the pond environment too: willows, sycamore, water oak, water tupelo, and swamp cottonwood.

Certain animals are regularly found at ponds. Mammals such as beaver, muskrat, swamp rabbit, rice rat, star-nosed mole, water shrew, meadow vole, and bat are often seen there.

Insects are numerous and varied. Water striders and boatmen scoot across the pond's surface. Dragonflies and damselflies streak overhead. Mayflies, caddisflies, fishflies, and beetles are everywhere. The larval forms of insects are an important food item of crayfish, salamanders, frogs, and fish. All of these, in turn, are eaten by turtles, the chief predator of which is often the wily raccoon.

The survival of this dynamic, ever-changing pond society rests on a base of tiny life forms, many of them visible only under a microscope. But others, such as insect larvae, copepods, and newly-hatched crustaceans can be studied with a scout's magnifying glass. I encourage my nephew to carry such a lens, plus a notebook and containers for specimens, when visiting ponds.

Moss animals (*Bryozoa*) with colonies that live anchored to submerged logs and rocks, proved fascinating under a bit of magnification. When feeding, there are thousands of tiny tentacles busy capturing algae, protozoans, and bits of decaying matter that drift by. But when the colony is disturbed, the tentacles abruptly retract.

Youngsters are also interested in the differing modes of water creature locomotion. Fish swim, of course. But *hydras*, those floating sac-like bodies fringed with tentacles, will somersault end-over-end through the water.

They are usually less than an inch long. Leeches, on the other hand, travel by looping their slender bodies forward as a caterpillar does, and are in fact classified as water worms. Fairy shrimp, less than an inch long, are even more bizarre in motion. They swim on their backs, bellies turned toward the sky.

Some ponds have a strongly individual character due to a unique feature that is remembered long after a visit.

One pond in the South, I recall vividly, was dominated by a solitary female water turkey (*Anhinga*) whose humped blackish body could usually be seen perched in a dead tree, wings outstretched like the American eagle on a dollar bill, waiting for her drenched feathers to dry between dives.

Another pond was terrorized by a giant snapping turtle that nearly depleted the fish population and would even boldly chase humans. The reign of this granddaddy of all snapping turtles ended at last in a bloody battle with a courageous coon hound.

One pond I remember fondly from my own childhood. A sizable mulberry tree leaned far out over its waters. The tree attracted birds of all kinds that dined on its gradually ripening fruits. Berries that dropped were eaten eagerly by the fish below. I picked quite a few of the berries myself. Mulberries are very sweet and not at all tart, but no child would object to that.

Both fish and ducks taken from that pond were fat, with a delicious flavor to their flesh, which may well have been due to the old mulberry tree.

Cooking a shore dinner of things from the pond is an experience nearly all children love, especially if they have helped gather and prepare the foods. Crayfish make a princely meal. Batter-fried froglegs or pan fish, however small, are winners. Cattail pollen can be gathered by the cupful in paper bags; added generously to pancake flour, it imparts a nutty taste and golden color, besides being high in protein. Fleshy cattail tubers and those of arrowhead are good roasted in ashes. Sprouts of cattail, boiled, are much like asparagus. Watercress and sheep sorrel make good salad herbs.

Some of the pancake flour should be saved to thicken into a pudding whatever berries or small fruits are nearby for the kids to pick. Sweetened with honey and perked up with a dash of lemon juice, even choke-cherries or the usually bland elderberries, which grow near water, can make tasty desserts.

Maybe anything tastes wonderful beside a campfire, but that doesn't detract from the pleasures of a wild banquet.

Surely there is no better way to stimulate youngsters to learn about nature than to share with them the enjoyment of its bounties beside a scenic pond.

JUNE

16

BATS: HUNTERS OF THE TWILIGHT

Missouri is cave country. And where there are caves, there are bats.

Some of the species found here are the Gray Bat, the Indiana Bat, Keen's Bat, the Silver-haired Bat, the Eastern Pipistrelle, the Big Brown Bat, the Little Brown Bat—and the Least Bat.

All these are shy, nocturnal, and harmless. They are insect eaters, and that is just what the bats are up to on a summer evening as they circle around some yard light or street lamp in swift, zigzag flight. They are not, as many believe, allured by the light itself, but are feeding on moths and other flying insects that are so attracted.

Bats are efficient feeders, said to be capable of eating an insect every few seconds, filling their stomachs in a half hour, and eating half their body weight in a single night. So this humble and often despised animal is, in fact, a farmer's good friend.

One neighbor tells me that as a boy he used to toss pebbles into the air by an outdoor light and watch the bats dive for them as eagerly as Mexican lads dive for coins dropped in Acapulco Bay. He denies ever breaking the light bulb, of course; he says he had better aim than that. The bats, he noticed, tried so hard to catch the pebbles in midair that often they seemed in danger of dashing themselves against the earth.

A Conservation Department man today, this neighbor

is far from approving of such sports now. But he tells the story to illustrate the zeal with which the bats hunted for what they took to be giant mosquitoes.

That bats have a sinister reputation is due mainly to the bad press they have received—occult tales, Dracula's legend, witchcraft, the supernatural, and vampire bats. There is indeed a bloodsucking bat of the tropics, although fruit and insect eating bats are common there also. But to malign our useful and harmless species of the Ozarks because of a distant relative or folk myths is as silly as being afraid of beagles because the dog family includes the timber wolves of the arctic. Or because of widespread werewolf beliefs. As for the occasional bat afflicted with rabies, that disease also can be carried by dogs, squirrels, and other mammals.

The most unpleasant thing involving bats, in my opinion, is their special relationship with the bedbug.

Here, too, the bat is more victim than culprit. The bedbug, that bane of boarding houses and roadside inns of the colonial era, is thought by scientists to have first evolved in Old World caves as a parasite on bats, turning its attention to humans only when bands of hunters began sharing the cave shelters in prehistoric times.

Most Missouri bats belong to the mouse-eared group. During winter they hibernate, choosing the warmer, drier sections of caverns. They hang upside down by their clawed hind feet from cave ceilings like hundreds of small globular stalactites.

From their deep sleep they may rouse briefly during warm spells to mate and lick moisture from cave walls—but not to eat.

When spring comes the females leave the caves first, forming into summer colonies to bear their young undisturbed. The males, leaving later, remain apart.

It is not surprising to learn that this aerial acrobat is born in midair. The mother bat, hanging head down from the cave ceiling, spreads her tail membrane to catch the emerging infant. She cleanses it and carries it with her everywhere for several days. Bat youngsters grow so fast they can fly and catch their own food at about four weeks.

During the long summer the bats sleep by day, roosting in thickly shaded trees, under loose bark, in caves and rocky crevices. They may also roost in barns, house attics, or steeples.

I was fortunate in finding a dozen Little Brown Bats (*Myotis lucifugus*) roosting under an old stone bridge, where I was able to examine them at leisure. Their fur was dark brown, long, and slightly wavy over the back, while the wings were largely bare and membranous. Their ears were narrow, with rounded tips. One surprising feature was the round puppylike face with its high domed, bulging forehead, far-apart eyes, blunt muzzle, and little pug nose, which reminded me of a Pekinese dog.

Here again is an example of what a bad press can do to an animal. The gargoyle face photographers love to use in illustrating magazine articles on bats is very often that of an ugly fellow called the Western Lump-Nosed Bat, usually pictured with its mouth wide open, showing all those sharp little fangs. But Lump-Nose, too, is an insect eater, and the gaping mouth emits supersonic squeaks that help locate the prey.

Maybe if the word "bat" called to mind those sleeping puppy faces instead, few people would fear the creatures.

At dusk in the summer, bats emerge from their hiding places to feed, and they are most often visible at this time. They do not coast on air currents as birds do, but fly by constantly beating their wings.

The erratic flight patterns of bats have always seemed mysterious. Only in modern times have researchers discovered the bat's peculiar sense organs and unique echo-location system. As it flies it continually emits ultra-sonic squeaks, experts say as many as thirty a second, rising up to fifty a second when approaching and passing an object. The bat orients itself by the returning sound waves or echoes.

In their own habitat, as when spiraling up by the hundreds from some cave exit into an evening sky, bats present a graceful, even awesome sight. But a lone bat trapped in a closed room, pursued by threatening humans, its weak eyes blinded by sudden lights, appears clumsy and pathetic.

One rainy morning last summer I chanced to be in a hall where a 4-H display was being readied for the county fair. When overhead lights were switched on, a bat fluttered out of a window drape. Instantly, all was wild confusion. Children screamed. Several panicky voices cried "Kill it! Kill it!"

The wife of the conservationist was calling just as loudly, "No, don't hurt it! Bats are a protected species. My husband says. . ."

The bat flapped overhead several times, trying to reach the door. But a well intentioned and sadly misguided youth had stationed himself there. Whipping off his shirt, he tried to net the bat each time it approached.

Impasse. The kids screamed louder.

Then someone with good sense opened the big double doors at the end of the hall. The bat slipped out and vanished in a maple tree.

Mark Twain, familiar with bats from frequent boyhood visits to the great cave below Hannibal, once said, "A bat is beautifully soft and silky; I do not know any creature that is pleasanter to touch or is more grateful for caressings, if offered in the right spirit."

Now I don't feel quite that cozy about bats or any other wild animal. Still, if a bat gets into my house, I just open the door, turn on bright lights (the bat will retreat from a lighter to a darker zone), and let the creature go back outside to the garden.

Who knows? It may repay the favor by eating half its weight in bugs that very night.

17

A PUFF ADDER PLAYS HIS TRICKS

I had just come out of the woods into bright sunlight with an armful of poles for tomato stakes when I spotted the kittens behaving oddly. Near the remains of last winter's woodpile the feline pair—named Smith and Wesson by my young nephew—were leaping about in a wide circle intent on something in the dead leaves.

Their wary manner said clearly, "We have something dangerous cornered here." The woodpile. Something hidden in dead leaves. The kittens' excitement, the wideness of their circle. These all added up to one thing—a snake. A big one!

Dropping my load, I seized one of the stakes and charged to the rescue. The snake was not, as I expected, a copperhead. It was rather short but heavy and thick-bodied, with a large blunt head. Its back was grayish brown with nondescript black markings. Coiled among dried leaves, it was well camouflaged.

Suddenly, as I hefted the tomato stake dubiously (it seemed all too short now) the snake whipped into an erect, striking position. Aghast, I saw the region behind the ugly head swell out into an awesome hood like a cobra's.

With a hiss like a steam radiator, the creature spat. Involuntarily I flinched, having heard that the saliva of some snakes can blind one's eyes. But no drops of moisture reached me. I raised the stake, prepared to stab. At the same instant Smith, the female kitten, leaped.

She sailed directly over the swaying head. One raking hind foot seemed barely to touch that scaly neck, but the hood shrank like a deflated balloon. The snake collapsed into inertness, its pale belly turned up.

It was all over. The kittens sniffed and batted at the motionless body. Then they grew bored with the lack of response and ambled away. Using my pole I tried to turn the snake over, to see the extent of the neck wound—surely the cat's claws must have severed the spinal column. But always the body slithered limply into belly-up position. It was rather like trying to straighten a plateful of wet noodles with one chopstick. I could not even study the coat pattern, although I did notice that no long fangs protruded from the gaping jaws.

In the house, I began going through my nature books, hoping to identify this mysterious snake with the cobra-type hood. Nothing quite fit his appearance, so I went outside to take another look at the markings.

The snake was gone without a trace. Not even a tiny spot of blood marked the site of its supposedly mortal wounding.

Disappointed, I returned to the books. This time I didn't just look at colored photos, I read about habits too. And I learned about the peculiar creature *Heterodon platyrhinos*, known also as the puff adder or the hog-nosed snake, which has three tricks. When disturbed it draws itself erect in a threatening position, spreading a hood behind its ears. Then it hisses explosively. If these two bluffs fail, it plays dead and will even allow itself to be handled. Then, like the opossum, it comes back to life and escapes at the first chance.

I had been fooled by all three tricks!

Neighbors to whom I related this incident said that in the Ozarks such a snake was called a spreadhead or puffer. "Harmless," said one farmer, "you can leave them around. They eat rats."

Since the hog-nosed snake was also listed as benign, without poison sacs, living on rodents and toads and insects, my identification of the mystery snake seemed correct.

Quickly I decided to coexist with Spreadhead Sam, as I named him, for the sake of maintaining ecological balance in my hollow. So I did not level the old woodpile, nor did I chase after the snake on its hunting rounds. Even the cats obliged by ignoring Sam as a dull fellow who was no fun to play with (unlike mice and moles).

In time, though, the snake moved to a large brushpile on the ridge not far from the fence of my neighbor's pasture. This high meadow abounds in mice, moles, shrews, and insects that hurl themselves in all directions as someone walks through the grass. So it is possible that a decreasing food supply down in the hollow was the reason for Sam's leaving his burrow under the woodpile.

Smith had become such a splendid mouser that she had eaten a poor hardworking reptile out of house and home!

Inevitably I hit on the idea of photographing this unique animal posed in the full glory of his bluff—erect, hooded, beady eyed, malignant of aspect. For a while I stalked him patiently, camera in hand. However, he seemed to have dropped the first two parts of his act, so whenever I approached him he would go directly into the big death scene, flopping belly up. That kind of picture has little appeal.

Probably he thought that his first two tricks didn't scare me at all on our original encounter. He was wrong, though. When I first started leafing through books, looking for a color plate of an American cobra, my mouth was still as dry as desert sands.

18

THE WOODS AFTER DARK

Being in the woods at dusk reveals a different world from the same spot on a summer day. New scents and strange sounds often defy identification. As some flowers only open in the dark, so a whole segment of the wild population of my hollow awakens in the cool of the evening.

Each kind of animal has its own reason for being abroad under the cloak of darkness. The reasons are as varied as the life forms themselves, but all have to do with one vital issue—survival.

The most visible of the night creatures, almost a symbol, is the firefly. Not a true fly but a slim, soft-bodied beetle with light organs on the abdomen, this insect is able to locate a mate only after the sun goes down. (In nature, only individuals who leave offspring have truly survived.) As he flits with his magic lamp among the trees, the firefly gives off intermittent flashes coded to attract a female of his own species.

One firefly writes fiery letter J's in the air. A second species moves in lazy S curves lighting up briefly at each apex, while another uses a three-dots-and-a-pause sequence. The system works amazingly well.

Almost always, that is, Here comes a tiny J-writer male now, excitedly homing in on an answering, correct signal from a clump of grass. How can he know that his particular code has been cracked by enemies? Instead of his rightful bride a large hungry *Photuris* beetle, which also possesses controlled luminosity, may be waiting there to devour him.

The most audible insect of the summer night, the katydid, is also courting. The male katydid is the musician. By scraping the rasps and ridges at the base of his wingcase he produces a mating call that, again, is peculiar to his species and will summon a female of his own kind.

His main problem is not that some predator will intercept the call and pounce on him, but that there may be no answer. For some reason males seem to outnumber the interested, fertile females in that society. This patient suitor never resigns himself to bachelorhood, but continues to scrape away at his wingcase fiddle all summer long.

Another class of night creature, like the legendary vampire, fears the sun. The rattlesnake hunts after dark because he cannot risk being long in broiling sunlight which, with his peculiar metabolism, might be fatal. He is comfortable and most active in temperatures between 70 and 80 degrees Fahrenheit.

With the rattlesnake may be included those small animals whose skin must remain moist: hop toads, lizards, snails, spring peepers, and the little tree frogs that chorus so vigorously in the moist night air. These do not avoid the light or heat of the sun but the resulting drying out of their tissues during such exposure.

A more numerous group of wildlife nocturnals includes the small, timid and defenseless, who hope to be able to feed in the dark unnoticed by predators. Mice, rabbits, opossums, and flying squirrels venture forth at night with ears cocked and noses twitching as they sniff the breeze for enemies. Yet they will sometimes ignore the beam of a flashlight turned on them and continue feeding.

White-tailed deer also prefer to browse at night, seeking out a thicket or secluded nook before sunrise, in which to pass the day.

The timid are most often vegetarian browsers, although some may include insects and scavengings in their diet. They have elected a nocturnal existence to avoid enemies, and to this end have developed specialties which include enlarged ears for keener hearing, big eyes that make the most of the faintest starlight, and an excellent sense of smell.

Their hunters, on the other hand, are often not specifically nocturnal. The bobcat, skunk, raccoon, fox, coyote, wolf, and bear can forage or mate just as well by daylight. It is primarily to avoid humankind that this predator group has been forced, in many regions, to be active at night.

Unlike the silent browsers, the meat eaters are a vocal bunch. Members of the raccoon family call constantly to one another—soft musical cries—as they prowl the night. The coyote pack's eerie, quavering hunting calls are well-known, and always a bit disturbing to human ears. The mating cry of a lovesick bobcat has been compared to a woman's scream, but to me it sounds like nothing on earth.

These sounds, like the barking of the fox, are identifiable voices. But in any woods at night there are also mysterious sounds, elusive voices. Flutings, whistles, and sometimes a feral shriek so appalling it makes those who hear it ask, "Now what was *that*?"

(I tell guests it's Bigfoot who has just found an interesting spoor.)

Night birds are a noisy bunch, too. Whippoorwills conduct all their business in the dark: feeding, courting, care of fledglings. Now bird calls are ordinarily associated with territorial claims, but I have never heard a good explanation of why whippoorwills chant in unison every night all summer, until even a second brood of young must be on the wing if not ready for migration.

The owl clan's vocalizations, on the other hand, are fitful and unexpected; they are clearly hunting calls. Truly nocturnal, owls show interesting adaptations that help them to operate in darkness. Those enormous eyes revealed in a flashlight's beam are specialized for seeing in the blackest night. Even the owls' hearing is said to be three-dimensional. Finally, its feathers are constructed so as not to give any warning rustles as it dives toward an indiscreet rodent.

Bats, completely nocturnal animals too, have good night vision but operate most efficiently with their incredible sonarlike echo-location system as they chase and catch insects in midair. Sometimes the intended victim has its own survival gift from Ma Nature. The lovely green Luna moth, for example, has a sensory organ that gives an alert whenever a bat's echo-location system is aimed in the moth's direction, so it knows just when to zigzag and hide.

Hunters and prey. Bold courtship and secret nibblings. All are part of the fascinating nightlife of Flatrock Hollow.

Right now it's past three o'clock in the morning, and every citizen of this remote woods is skillfully pursuing his own business. Except for one who, sitting on the doorstep, does nothing but watch and listen. For this individual is actually a day creature that, having gotten into a habit of going to bed with the chickens, has already had a full night's sleep and is waiting for the dawn.

And thinking of putting on the old coffeepot.

JULY

19

WHO'S BEEN PICKING IN MY BERRY PATCH?

My first day of blackberry picking in the Ozarks was a bit unsettling.

For one thing, the sheer size of my neighbor's wild berry patch, where I had been invited to help harvest the crop, was staggering. No backyard fence row, this! It spread across a rolling twenty-acre corner cow pasture. Also, these Ozark berry bushes were the largest I had ever seen, with canes towering high over our heads.

The blackberries were huge too—at least that year—as big as my thumb. I ate a couple of ripe ones and found they were delicious.

The trouble was that of the thousands of clustered berries in the sea of bushes almost all were bright red. Only a few scattered dark ripe ones were to be seen.

"Too early in the season," I sighed, disappointed. "Let's come back in two or three days."

"Come on, wade in!" Seeming not to have heard me, my hostess, wife of a retired railroad man from Denver, was already moving from bush to bush, garnering the meager harvest of dark fruit. I followed suit and soon the bottom of my (overlarge) pail was covered with a layer of choice berries.

In less than an hour the two of us circled the entire berry patch. We had about a quart and a half of berries apiece.

"Fresh blackberry pie tonight!" My friend was

pleased. Then her face darkened. "But there should have been more ripe berries today. Somebody's been picking in my patch. It's too near the road. Wish I could catch the thieves just once. But I can't be watching out the windows all day. Probably just some kids, anyway."

Later, as I washed those gigantic berries—enough for a big pie and a good portion left for freezing—I thought ruefully of the berry picking expeditions of my youth. The brimming buckets, the gallons of juice boiling down for jelly on the old woodstove. I missed the wine crocks and their yeasty smell.

Another thing I missed was the scrambled egg sandwiches.

Always, when I was a child, the evening before blackberry picking my dad would look around the supper table in that old Pennsylvania farm kitchen and say in a firm voice, "Tomorrow we'll get those berries in."

But that was not the decisive moment. The whole thing might still be called off by bad weather or sudden illness. The exciting moment when we kids really knew we were going, and felt our hearts beat faster in anticipation of the treat, was the next morning while it was still dark outside, and Mother began breaking a dozen eggs into the big iron skillet to scramble for our sandwiches.

Berry picking in the old days!

Carrying all the zinc buckets and big kettles in the house, we followed Dad across meadows and into the woods, soon falling into Indian file for easier walking. In the east the sky would be reddening with dawn. Somewhere a farm dog would bark, and a bird nearby would chirp sleepily in answer. At the edge of the deeper forest, blue-shadowed aisles beckoned. Swinging empty pails, we kept going. Any blackberries growing along roads, fences, or at the edge of the woods we left behind for the townfolk to come out and pick. We were headed for the really choice spots where the biggest berries were, like the clearfields.

Clearfields were sunny glades in the deep woods, where no big trees encroached on the circle. Our new teacher at school called them fairy ballrooms. But we were one of the early settler families and knew them to be old Indian burial grounds. I don't know why the berries grew better there.

The long brush-choked swath under the powerlines was another good berrying place, but often the best of all was some hillside slope where the logging crews had worked a couple of years before.

When we reached a good patch our holiday spirits were reined in by meaningful throat clearing from Dad. And soon the only noise to be heard was the heavy drumming of berries in our pails, a sound like rain on the roof.

We picked into two-quart coffee cans with holes punched under the top rim for cords so the can could be slung from a shoulder, freeing both hands for picking. The quick motion used was reminiscent of milking cows as we went from bush to bush, reaching always for more berries hanging just ahead. When the coffee cans were full, we emptied them into one of the larger containers, and there was keen rivalry to see who could fill a pail most quickly.

While picking clean, that is. We were disqualified if we ended up with twigs, leaves, and stinkbugs in our berries as a result of hurrying.

It was backbreaking work, though, trudging from one patch to the next, filling pail after pail, reaching, squatting, toting burdens that grew ever heavier as we stumbled through briars and over logs.

Still, we had fun. And did anything ever taste as good,

in the heat of day, as those buttery scrambled egg sandwiches on generous slabs of homemade bread, with cold spring water and all the berries we could eat!

That evening there would be flaky crusted deep-dish blackberry pie for supper, served hot in bowls so we could pour milk over it. And then to bed, early. But not to sleep. Or at least, not restful sleep. In a special state of fatigue of mind and eyes as well as body, we kept seeing all those berries hanging before us. Endlessly our cramped fingers reached for them, twitching involuntarily in a milking motion. The vision persisted whether our dazed eyes were open or shut. Masses of berries danced on our closed eyelids or took form in the darkened corners of the ceiling.

When at last we slept, all night we dreamed of picking berries, and in our ears was the sound of berries drumming on the bottom of the pail, echoing like steady rain on the roof.

Yes, that one thing was missing and not regretted in my Ozark berry-picking experience. A few chigger bites, maybe, but no berry picker's daze afterward to ruin my sleep.

In my memories of the old days, also, all the berries were ripe at once. If we came picking at the right time, that is. A day or two earlier, and the berries were still red. A day too late, the patch was stripped bare—someone else had found it, the birds had fed heavily, or rain had caused the berries to drop off and be lost among sodden leaves on the ground.

But what a harvest for those who kept their eyes on the berries, as my dad did. He could tell, even at the time the white starlike blossoms opened in the spring, which patches would be worth visiting in fruiting season. In a good year my mother might put up eighty quarts of berries, plus six dozen jars of jelly and a few bottles of sparkling dark red wine—a mere sip of which was guaranteed to cure a stomachache or cold.

Curiously, in the Ozarks this exact right day for picking berries never comes. Or rather, almost any day is all right. That first year I was invited over several times, by the retired couple, at three-day intervals. Each time was like the first. Only a few ripe berries. A quart or two per person, as a rule. At the end of each picking session, too, came the apologetic but half angry remark, "There should have been more ripe ones today. Somebody's been picking in my berry patch."

Who were the berry thieves? None of the other neighbors, who had their own big berry patches, and all believed they were being raided too. Watches were kept sporadically, but the stealthy intruders were never caught.

Only gradually did I figure it out. Climate differential. The climate of the Ozarks is semitropical compared with that of Pennsylvania, where winters are long and frost strikes late in the spring, early in autumn. Up north, the plant kingdom is in a hurry to blossom, fruit, and seed out in a short growing season. That's why the whole crop ripens at once. But in Missouri, Nature spreads the process out over weeks. And down in the real tropics one finds blossoms, green and ripe fruit on the same tree all year long.

Very soon the leisurely Ozark season suited me just fine. For a middle-aged person it's great to be able to put off strenuous deeds until tomorrow or even the day after. That just lets more fruit get ripe for the next trip out. I enjoy taking my time, having all the fresh berries I can eat, and letting the surplus collect in the freezer for that day I feel chipper enough to make up a batch of jelly. No rush, no spoilage, no aching legs from hiking miles over rough terrain in one day. In the Ozarks the berries stay fresh, and I can always find more ripe ones waiting on the bush.

Unless somebody else gets there first.

Not only did I not convince my neighbors that there was no berry thief—only climate differential—but after a while I myself began to doubt that it was the whole story.

That was after I got my own private berry patch, on a secluded acre where I had cut firewood a few years earlier. And not only are there far fewer berries than one could reasonably expect, there are signs—canes freshly broken and new trails trampled out all over the patch with reckless disregard for the tender shoots that will bear next year's fruit. Since it's my own patch I see these things.

At last, someone has identified the berry thief. The retired couple from Denver just phoned, elated. It seems that their close neighbor, driving past the berry patch in his truck around daybreak, spotted a small brown bear there that is thought to have its den somewhere down in Bone Creek Hollow.

This bear was loitering under an oak at the edge of the pasture, licking one paw. Obviously he had just left off berry picking.

Everybody knows that bears love berries and honey.

Curiously enough my friends, who would have been furious had the thief been human, were as pleased as could be about the news.

Well, from now on the brown bear of Bone Creek can have all the berries he wants from my patch too.

20

A CAVE IS A WILDERNESS
AREA

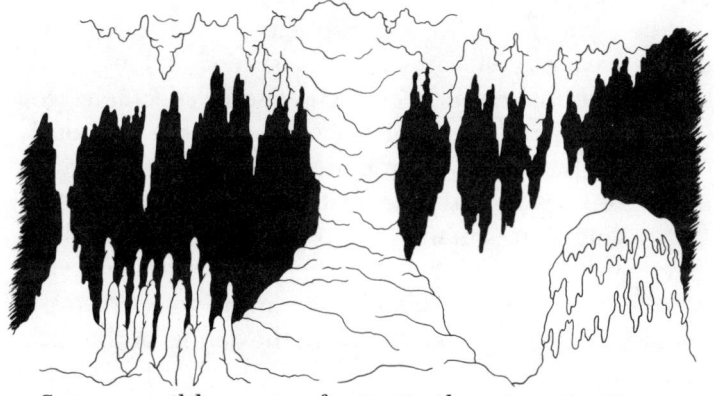

Scenery wilder, more fantastic than imagination can
conjure, tinged with the romance of history. Dreams of
golden hoards hidden by outlaws and river pirates. The
scientific study of rare life forms. Strenuous physical exercise with the sheer excitement and peril of mountain
climbing. These are part of cave exploration.

For hazards, caving—or spelunking—has been compared to mountain climbing on a black moonless night.

The experienced explorer uses the same skills as the
mountaineer and much the same equipment. Ropes,
pitons, spikes, and even a lightweight collapsible ladder
of steel cable and aluminum come in handy. The
"floor" of a cavern system is seldom as smooth and flat
as those in the movies. There may be huge mounds of
rubble from fallen ceilings, stark chimneys to climb,
gaping chasms to bridge, shadowy pits to descend.

The broad corridor you follow may constrict to a

small tube along which you crawl on hands and knees, while projecting rocks of the ceiling scrape at your back. Next you emerge from the tube on a narrow ledge fifty feet up on the wall of a vast amphitheater the size of a cathedral.

Such contrasts and unexpected vistas are part of the joys of caving.

When you are with a group exploring a newly discovered cave, you can be fairly sure that few if any human eyes have ever seen the awesome spectacles you encounter there. Being the very first to see a previously unseen surface area—even in the most remote deserts or mountains—is highly improbable.

If caving shares many features with mountain climbing, some of the latter's dangers are missing. Foul weather, for instance. The subterranean temperature is about fifty degrees year round, in the depths of the cave, so you'll never get frostbite there. No gale force wind will tear you from a high pinnacle. No driving snow or fog will blind you or cause you to hole up in your tent for days at a time.

On the other hand, if you get into trouble down below, no helicopter can come to your rescue. No airplane searcher will spot your distress signals. You can't even radio for help from a cave. As a spelunker you are definitely on your own resources.

Therefore, safe cave exploration has a set of rules to be rigidly kept. Some of the more important are as follows: Never go into a cave alone. Groups of three or more are traditional. Take the proper equipment, including nylon rope, compass for directions, pocket knife, and a first-aid kit. Wear a good wristwatch, because you can lose all sense of time down there. Take luminous tape for marking trail. Wear sturdy hiking boots, suitable for rough terrain, and warm clothes (a temperature of fifty degrees plus the dampness of lower levels can be bone chilling after a couple of hours, and hypothermia is a serious risk). Thick rubber pads to be strapped about your knees can be an asset on long crawls.

Always have three kinds of light: a carbide lamp (affixed to a hard hat such as miners wear to protect your head from falling debris) will leave your hands free for climbing; a good flashlight with spare batteries; and a couple of candles and matches.

Take at least a canteen of water. Even if you find pools, waterfalls, and underground rivers, the water may be undrinkable because of the acrid minerals dissolved in it. Also, farmers may have tossed dead cattle into sinkholes, polluting cave waters beneath.

Besides regular mountain climber's equipment, scuba gear and even a light boat are useful if you have found an underground lake. This combines yet another exhilarating sport with caving.

As a sport, though, caving has some hazards that are peculiarly its own. It is risky to shout, sing loudly, or fire off a gun in a cave. Reverberating sound waves could bring down the roof. Another risk is bad air. This is rare in natural caves, but if you suspect the chamber you're entering may contain a pocket of methane gas or carbon dioxide, light a candle and see if it will stay lit. On the other hand, never strike a match around deposits of bat guano. It is rich in potassium nitrate, an active ingredient of gunpowder, and is highly explosive.

Finally, always check the weather forecasts topside when planning a venture below ground. During a snow melt or after a rainstorm, waters may swirl down into the caves, trapping explorers in a flash flood.

What is there to be found in caves that makes it worth the risks?

Meeting the sporting challenge, for one thing. The magnificent scenery, for another. In a room that has been abysmally dark for unguessed centuries, your carbide lamp may reveal walls and ceilings of crystal that sparkle like the heaped treasures of Ali Baba. Violet shaded grottos hold pools of pale "moon-milk" fringed with the delicate flowers of helictite. Amphitheaters big enough for a giant's ballroom may contain monstrous pillarlike stalactites and stalagmites. Colors can be as rich as those of the Grand Canyon. Massive flowstones that look like carved waterfalls, bejeweled curtains behind ebony thrones, huge organs with alabaster pipes, fairy palaces and castles with translucent walls—the imagination can have a heyday in caverns.

Cave life can be fascinating, too.

Some of the cave creatures are as unique as their habitat. Only a few kinds of plants grow there—mushrooms, fungi, and molds, some of them luminous. But over a hundred kinds of animals have been known to live in caves. Besides the familiar bears, cougars, foxes, rats, snakes, and the like that make their dens near the entances, and the hordes of bats that penetrate deeper to nest in secluded chambers, there are the true cave dwellers or troglodytes.

Troglodytes are life forms adapted to survival in perpetual darkness. They include white salamanders, fish, lizards, crayfish, crickets, spiders, flatworms, and ants, blind because sight is useless here. Their bodies are ghostly pale, white, and some nearly transparent. One glasslike species of blindfish (*Amblyopsis*) was discovered when an alert caver wondered what was casting the shadows of fish he noticed moving on the bottom of a cave pool.

Missouri is justly called the cave fancier's paradise. It

has more caves reported and open to the public than any other state. You can visit the cave Daniel Boone discovered on his travels, the one that Jesse James and his gang hid in with their loot when hotly pursued by the sheriff's posse, the one Mark Twain loved and described in his adventures of Tom Sawyer, and others rich in historical association.

Perhaps as many more caves here are unexplored or undiscovered, awaiting the curious person who notices a cool draft of air coming from beneath a humped tree root, or sees a bird fly into a dark hole on a bluff, or investigates a sinkhole that suddenly appeared in a pasture. Creeks that vanish underground are another dead giveaway.

Even though the enchantments of a cavern are sculpted in stone, the beauty is perishable, easily mutilated, and the rare species of wildlife are extremely vulnerable.

Caves open to tourists are usually guarded by their managers against defacing and thoughtless vandalism. In a sense, such places are national monuments, like the ones above ground—a heritage that should be protected.

It follows that a newly discovered cavern system, even if the access is on private land, should be treated as a true wilderness area. If you're lucky enough to find one, act as if you were the first person to visit the Painted Desert or to stand on top a previously unclimbed peak. No carving cynical graffiti on the cave walls. No breaking off crystals or small stalactites as souvenirs for friends. No taking home a whitefish for the kids' acquarium. And, of course, no tossing candy wrappers or cigarette butts.

As the saying goes, enjoy the sights. Take nothing but pictures; leave nothing but footprints in these magical marbled corridors beneath the weathered Ozark hills.

21

TRACKING THE RAIN CROW

"Listen!" My neighbor raised one hand. He was standing in his dusty beanfield where a few grasshoppers moved sluggishly. Even a clump of foxtail grass under the fence was languishing in the heat.

Nothing was to be heard but the soughing of the west wind, and from a nearby wooded hill, the sharp cry of a bird. Not a melodious sound, it was a strange guttural squall: "*Kok-kok-kok-kok-kok-kok. Kowp-kowp-kowp.*" The last few notes came slower and slower, like an old music box running down.

"It's good to hear that sound," the farmer said, looking pleased.

"What is it?" I asked, being new to the Ozarks then.

"Why, the rain crow. When it calls that way, it means the end of the drought. Rain's coming!"

Sure enough, we had a real downpour within hours.

After that I grew curious to see this feathered prophet of the hills.

For weeks I haunted the wooded slope from which the raucous cries had come. From time to time I heard the bird calling, but never caught a glimpse of it.

I knew what crows looked like, of course—big ugly flapping black things. As a kid I'd seen a dead crow hung on a farmer's fence post from time to time. A check of my handy bird guide's section on crows and ravens showed that they hadn't changed their looks any. None of the species listed there were identified as *the* rain crow either, which was a bit discouraging.

A couple of years later I was set on the right track by an old timer who remarked that the rain crow builds about the sloppiest nest of any bird.

Back to the wooded slope I went with this clue, looking for nests. As a rule I have no luck at spotting them from the ground until the leaves fall off the trees, by which time it is too late. But I kept at it. I got a crick in my neck from looking up and bruised shins from not watching where I stepped. But the nest proved as elusive as the bird.

Then came a real stroke of luck. Actually I was gazing down when I realized a nest was near.

Fragments of blue-green eggshell were scattered about on the ground. Just then another small piece dropped from above. Overhead in a thorn tree, and not eight feet from the ground, was the poorest excuse for a nest I have ever seen. It was a mere platform of twigs and other material, so carelessly assembled that two other blue-green eggs could be seen through the cracks. Also, something small and dark that was stirring in the nest.

Backtracking up the hill behind the thorn tree and

using binoculars, I could look down into the nest with no trouble.

The dark object in the nest was a newly hatched chick, ugly, nearly naked. Evidently his movements had knocked the pieces of discarded shell through the gaping holes in the bottom of the nest, and I held my breath for a moment when it seemed one of the eggs might follow suit. Happily, this tragedy did not occur.

Sloppy construction, ugly black chick. This had to be the rain crow's nest.

Elated, I retired behind a tall bush to await the arrival of the mother bird so unaccountably absent during a crisis like the hatching of her eggs.

When the mother arrived, I stared in amazement. She was a graceful dovelike creature with chestnut wings, white throat and belly, and a long black tail marked with dramatic white splotches.

The youngster hissed excitedly, and she answered with a raspy cluck. Something was wrong. The voice of such a pretty bird should be a restful coo, not a croak.

Hastily I leafed through the bird guide. And there was a picture of the bird. It was called the yellow-billed cuckoo (*Coccyzus americanus*). From the text I found it to be one of the few birds that will eat hairy caterpillars, dotes on webworms, and thus is the farmer's good friend.

This American cuckoo, I learned, is a careless nest builder. And although she does not parasitize other species as does the European cuckoo, she occasionally lays an egg in another bird's nest, which seemed absent-minded to me. At the bottom of the page was added,

"Also known popularly as the rain crow."

It made sense. We label as "cuckoo" a person who is forgetful, inept, or behaves peculiarly. Judging from the half-finished nest and other habits, our local weather prophet appeared to be just such a character.

Three days passed before the last of the other eggs had hatched. It seems that the rain crow begins incubating before she has finished laying her clutch. Coming from the shell, the youngsters are nearly naked but develop quills which open into true feathers in about a week. This latter change, according to one observer, goes as fast as popcorn popping. I missed out on this oddity but did see another startling event from behind my tall bush.

Evidently the youngsters didn't think much of their nest either, because they left home mighty early. Still incapable of flying, they began swarming all over that thorn tree, holding on with their claws. Maybe they were in hot pursuit of bugs to eat, I couldn't tell. But the parents remembered to feed them too whenever they were in the area, so this nest full grew up to look as handsome and dovelike as their elders.

When I told my little nephew about finding the rain crow's nest, he said "Great!"

Guessing he didn't mean my tracking ability, I asked, "That the little ones made it?"

"Well, that too," he said. "But I had an awful thought. What if someone shot the rain crow and hung it on a fence, thinking it was just an ordinary crow? Then we'd have nothing but hot, dry summers after that. But *nobody* would shoot a bird that looks like the dove of peace and eats hairy caterpillars!"

AUGUST

22

PROSPECTING ON A SMALL CREEK

The creek that flows past my door in Flat Rock Hollow is small, when not swollen by a sudden flood. Usually much of the stream bed will be dry a week after a rainstorm or snow melt, however heavy.

The ravine is not quite what Westerners call an arroyo, because here and there are rocky basins that will hold enough water for the needs of the wild animal population of this wooded hollow.

It is just at the stage when the water is running low, about four days after a rain, that I like to walk up the creek prospecting. In my own way, of course. Reaching an affinity with nature is as important to me as any treasure I might find, freshly washed out of the earth.

With newly disturbed pudding stones shifting under my boots, I hike along the creek at an hour when the rays of the morning sun glint off the damp sands, bringing out unexpected rainbow hues and golden sparkles from half-hidden pebbles.

Always I look for fresh changes wrought by the latest storm. Here the water, cutting across a projecting sandbar, has created a new island. There an undercut bank has collapsed to reveal an interesting cross-section of hillside: a thick layer of limestone over red-brown travertine, and above that a thin layer of red sandstone, topped by several inches of black soil smelling of leaf mold.

The earth of the bank is pierced by small, round holes, part of the abruptly exposed burrow of some animal.

Further on, a dead cottonwood has fallen diagonally across the creek.

The log makes a good place for me to sit, legs dangling, while I look into the shallow pool that each spring is alive with darting tadpoles. Since there are never any adult frogs to be seen here later, such tadpoles may grow up to be the little tree frogs (*Hyla*) whose elfin "reet-reetings" chime between the staccato raspings of katydids in the summer night.

Ambling along, I do a spot of prospecting. Now, when the sediments of the creek bed are freshly churned and new material is washed downstream, is the time to look for gold nuggets.

That's right, nuggets. This isn't the Rockies, with a legendary mother lode lurking somewhere, but gold is of almost universal distribution on this planet. Small amounts of it have been found in every land and in every state in the Union. So I keep my eyes open for that stray nugget.

Also, I watch for any golden *doubloons* or *louis d'or* that may have been dropped here by a passing explorer, centuries ago. The odds may be long, but what can it hurt to look?

It is slightly more realistic, in the Ozarks, to look for corundum. These six-sided crystals are often found in riverine deposits. Corundum is the hardest of all gems, save for the diamond, and is much used industrially. Colorless in its pure form, the crystal is not used in jewelry since no cutting can give it a sparkle, but some showing a six-pointed star inside the milky depths of the crystal are attractive.

If this stone has impurities of iron and titanium, however, it will be blue and is called a sapphire. If the impurity is chromium, the crystal will be red and is known as a ruby. Both of these stones are valuable

And then there's the diamond. The famous Star of Arkansas, found by a rock hound at Murfreesboro, Arkansas, weighed 40.23 carats and is probably worth $250,000 today. Diamonds are found in kimberlite, an igneous rock, and are often washed up in alluvial deposits—that is, by floods. Well, we get a lot of flooding around here, but thus far, no diamonds. (I look, anyway.)

What I do find in abundance when walking up any mountain creek are pretty colored pebbles. A good handbook on rocks and minerals can help to distinguish beryls, agate, tourmaline, spinel, rose quartz, or carnelian, all of which have some market value and make nice jewelry to give friends (especially those who are amateur rock hunters).

Old cultural artifacts are also found along streams—not just coins but arrowheads, spearpoints, and scrapers. Indian pottery shards may be embedded in the banks, too.

Like modern campers, Indian bands preferred to camp near streams or springs so the squaws would not have to carry water a great distance. I must admit that I have not found any discarded Indian pottery along Flat Rock Creek, but I did find a charming relic there from a different historical period. It is a rim piece with attached handle from the kind of little brown jug associated with the old-time moonshiners and their stills in the woods.

History—the long, slow sweep of time—is very much on my mind as I walk up the creek.

The Ozarks are old. These ancient mountains were weathered when the Appalachians were jagged young peaks, and the area of the present Rockies was a flat

117

coastal plain. In this secluded hollow, rushing flood waters through the ages have carved the ravine deep into the earth, revealing the rocky mantle that is the bare bone of the mountain. The wet stony ledges I climb, like a giant's staircase, represent successive layers or strata of rock built up, eroded, and rebuilt over aeons of time.

Some of these flat ledges show ripples, the wave pattern left by the tides of some ancient sea, frozen in rock.

Before there were mountains here, there was water. Ages before volcanic furnaces squeezed a chunk of carbon into the Star of Arkansas, the ocean was building up the thick layers of calcite or limestone that lie under Missouri soil. These were created a millimeter at a time by the piled-up skeletons of billions of tiny sea creatures.

On its retreat the sea left spongy-looking fossil-bearing rocks marked with flowerlike designs that were actually the work of coral polyps or marine worms. I select a nicely-shaped chunk to carry home for a paperweight.

Some living fossils can be seen here in the Ozarks, too.

That daintily branched dogwood is a survivor of the Redwood Floral Complex that overspread much of this continent between two Ice Ages. The opossum climbing a wild cherry is little changed from an ancestral form that emerged during the closing Age of Giant Reptiles. An even earlier period, predinosaur, is represented by an orange-necked turtle basking in the sun. And an incredibly archaic life form is seen in the green darner hovering over the water. Its wings are a primitive style that cannot be folded across the back like the wings of evolved insects like bees.

Almost always I find something worth carrying home—a specimen or a thought for my notebook—from these solitary casual prospecting trips.

My most unusual hike up the creek was in the company of my twelve-year-old nephew, who was visiting. I was sharing with him my philosophical musings on time and mountain building. The boy was walking a short way behind me, which made it very painful when he began finding treasures right and left, that I had strolled past unseeing.

Two nice flint arrowheads, from which he carefully brushed off all trace of mud with his shirt before handing to me. A tiny perfect bird point as white as alabaster. Some smooth pebbles of varied colors that looked good enough to be set in jewelry, just as they were. And I was really amazed at the handsome hunk of jasper, brown with bands of red and yellow, which he picked up in my very footsteps. How had I ever overlooked that?

If next he had found a gold nugget, I would have fainted. As it was, I resolved first to see an eye doctor about glasses, and second to keep my mind out of the clouds when young folk were around.

The treasures were duly spread out on the kitchen table and admired again. But my kid sister looked at her son suspiciously.

"Jimmy, I don't know about the rest of this stuff," she said, "but isn't that piece of jasper the one you bought in New Mexico? I recall you said it was your lucky stone, and you'd always carry it with you."

All at once my nephew was so weak with laughter he could hardly stand. (Too late, I remembered the trick he had played with a rubber spider on his last visit.) As it turned out, all the stones had been bought out West. As for the arrowheads, he had quietly purchased them at the gift counter of the Jesse James Museum only the day before, while I was showing his parents other exhibits.

Yes, I learned a good lesson that time. If ever I go

prospecting with another person, even a child, I'll watch out for claim-salting.

23

PAULA POSSUM, A WILD FRIEND

Dusk had settled over the Ozarks. Somewhere a barn owl hooted at intervals. There was a waning moon, the evening I first met Paula Possum. From the start we were friends. An odd friendship, based on mutual misunderstanding, but a solid and lasting one for all that.

Now I don't know what kind of animal the opossum thought I was—a long sinuous greenclad arm stretching down from above under a half-opened door, to set more food on the empty tin pie plate which she was pushing about on the concrete steps. But I thought she was another cat, when I refilled the pan and began to fondle her head. Strangely, this wild animal not only tolerated the caress but kept on eating with every sign of unconcern.

It was I who jerked back my hand, shocked on touching a mass of rough wiry hair instead of the smoothness of cat fur.

Eager to see what kind of creature the hairy one was, I stood up and swung the door wide, so that a shaft of lamplight spilled out across the front steps and lawn. At once the animal was off, with an odd lurching run. The lamplight revealed plump piglike hindquarters, a long hairless tail, and shaggy coat. A narrow white triangle of a face topped with rounded bat ears tilted backward over one shoulder as if to say, "Why do you chase me away? I was just cleaning up that mess."

Paul, the new tomcat, was an opossum.

For weeks I had been aware of a visitor sharing the food I left outside for Peter, my old black tomcat. Several evenings, glancing from a window, I had spied two animals sitting companionably opposite each other, eating from the same plate. First one would seize a bit of food, dragging it back onto the concrete to finish it, then the other took a turn. The black shape was Peter, of course. The stranger was light colored, maybe fawn or gray, and smaller. My reason for naming the newcomer Paul, and thinking it was a young male cat, was Peter's behavior.

Peter gallantly defers to females and kittens, allowing them to eat their fill before he even approaches the food, but he wades right in when fed with an adult male. Incredibly, now he had extended his courtesy to sharing a plate with a hungry young opossum.

Paul became a nightly visitor. The old pie plate by the door was only the main course. Unlike cats, opossums are not finicky eaters, and soon the stranger was raiding my garbage too. Peach peels, apple cores, fresh corn-cobs, chicken bones, fish heads, the trimmings of freezer-burned venison were all acceptable.

In the wilds, too, opossums dine on varied fare including eggs, fruit, insects, crayfish, and field mice. Unpopular with many folks because of its scavenging habits, the opossum is considered as inveterate a raider of chicken coops as the raccoon. My neighbors earnestly advised me to shoot the critter instead of feeding it.

My abode is a long way from any chicken coops, however, and one of my reasons for living in the woods is the peaceful enjoyment of nature, so I declined this advice. My interest awakened, I turned to books for information on the opossum. The name is derived from the Algonquian word *apasum*, a white animal. Nocturnal, largely arboreal, *Didelphus virginiana* has large eyes for seeing in the dark, a prehensile monkeylike tail, and separate digits that leave a mark in the snow like the palm print of a small hand. Also, there is a handy pouch for carrying the young. These are tiny, weighing about a fifteenth of an ounce at birth. One biologist remarks that you could pile a dozen of them into a tablespoon without undue crowding. Only about half of those born survive infancy, riding in their mother's abdominal pouch like little kangaroos.

The opossum is a living fossil, in a sense, and North America's only marsupial. Without natural weapons, slower and less brainy than the placental mammals that replaced their kind on this continent almost two hundred million years ago, the opossum has elected a stealthy nocturnal life to avoid predators.

Its sole defense, when attacked, is to pretend death— "playing possum," as the saying is. This works surprisingly well, as older opossums when examined often show large numbers of pelt scars, and healed fractures of many bones.

A living fossil? With some awe, then, I pursued my acquaintance with this young animal. For weeks after that first startling meeting I dared not open the door when my visitor was outside eating, but loaded the plate nightly and strained my eyes from a window.

Then Paul himself (for I still thought of the animal as male) took the initiative. This opossum was bright enough, it seemed, to adopt the tomcat's request for more chow, made by scraping the empty tin plate noisily on the concrete until the door opens a crack and more food is dropped. At such times I was again allowed to pat the little head and shaggy shoulders—but not to open the door widely enough to step out.

At the first hint of that, my shy visitor scuttled off into darkness.

On summer nights, when I sat quietly in a lawnchair watching the dance of the fireflies, often I saw Paul come out of the woods and trot in a businesslike way to what was now his food plate. (The tomcat had begun taking his meal earlier.) Even in bright moonlight, when he must have seen me clearly, he seemed undisturbed by my presence. If I moved, though, he fled.

It was early spring, when the trees were still bare, that I caught my first daylight glimpse of this creature of the night. On a sunny hillside above my house, two opossums were engaged in a courtship ritual. One was a lean handsome stranger of charcoal gray with long white guard hairs. He was ardently pursuing the plump, cream-colored animal I had been calling Paul, but who was now revealed as unmistakably female.

So absorbed were they that I watched unnoticed, less than forty feet away, the clumsy play of these two small survivors from a remote prehistoric age for whom, that day, the world was still young.

Evidently nothing resulted from that courtship. My next sight of Paula by daylight was not a happy one. One hot morning later in the summer, as I sat typing by a window, I was disturbed by the loud buzzing of flies outside—hundreds of flies, as in a Hitchcock horror movie. As I stepped outside, a black cloud of insects lifted from a furry bloodstained body slumped against the side of the house.

It was Paula. There was a gunshot wound in her flank and ugly mangling by dog teeth on one leg. The puffy, clotted wounds were hours old. Evidently she had been wounded during the night and had used her last strength to drag herself to my door.

There, with the rising of the sun, the flies had found her.

What disturbed me even more than the terrible injuries was the massive yellow cluster of flies' eggs the size of a child's fist, glued to the wound edges.

My foot scraped on a stone. The opossum's eyes opened. Lurching to her feet, the clumsy waddle exaggerated, she dodged back from me, around a tree, and disappeared into the darkness under the toolshed. With a flashlight I located her lying against a post, motionless again except for the occasional twitching of the wounded flank.

Mercifully the flies had not pursued her into the dark. But the damage was done. Bad as the idea was of the animal dying in slow agony, it was worse to think of those fly eggs hatching into maggots that would eat into living flesh. With a lump in my throat, I loaded a .22 pistol.

However, when I returned to peer under the shed again, gun in one hand and a flashlight in the other, it was to find that Paula had moved to a spot behind a post, where I could not get a clean shot at her.

That was fortunate. Because opossums, those living fossils, are very tenacious of life. They must possess excellent powers of wound healing and resistance to infection, to have survived in the dangerous era of fast, brainy, well-weaponed carnivores. I left her a pan of water and went indoors.

By midafternoon the little head was rising and falling regularly as she licked her wounds to clean them, and when next I got a good look, the monstrous egg mass was gone. I could not think they had hatched already. Most likely the practical Paula had eaten them. Well, why not? Insect larvae are eaten avidly by chickens, fish, and frogs, all of which are items of human diet.

Still, I knew what an old mountain woman meant when she offered to cook me a tasty meal of roast 'possum and sweet potatoes if I would bring her the live opossum and let her feed it on corn for a month first.

The space under the shed has proved a haven for Paula many times since. When coon hounds run through the hollow baying, or when winter snows are deep, she takes up residence there. But somewhere in the woods she has another home, a den, for once after a prolonged absence she came back one night with company.

Alerted by the scraping pie plate, I leaned out as usual to drop more food, and gaped in awe. There were guests—four of them. Bright eyes set in small white V-shaped faces, tiny hands grasping the fur on their mother's back as they were carried forth to explore the world. The smallest one hissed at me.

Although honored by this opossum mother's apparent trust in me, I do not feel she is becoming domesticated. Wisely, my wild friend (not my *pet*) has retained her independence, with the den in the woods and frequent absences. My doorstep with its magically refilling food pan is just one of the choice spots Paula showed her youngsters.

There are also mossy creek banks where crayfish lurk, places where turtle eggs can be dug, fencerows of persimmon trees, pastures rich in mice, and abandoned farms where wild plum thickets encroach on old apple orchards. Not to mention the little magnolia tree by a mobile home where a retired couple keep trustingly hanging up lumps of suet for the woodpeckers.

In all such places a living fossil can share with her young new tips as well as the age-old lore of survival. And tricks like playing possum.

24

RAID ON A WILD BEE TREE

The young couple that had taken a farm about three miles down the road from Dad's had kept to themselves for some months. But the man had a worried look when he came up to Dad one sultry day in late summer.

"You the fellow who has *bees*?" The way he said it made it seem a dirty word. When Dad admitted he kept a couple of hives, the man went on.

"There's wild bees in the woods at my place, in a hollow tree. They come right into my wife's flower garden now and drive her out. She got a sting on her temple the other day, and both her eyes swelled shut. The doctor said bees are poison to her. Now I'm not one to ask favors. If the woods wasn't dry as tinder I'd take a can of gasoline and burn the. . ."

"No, no," Dad broke in quickly. "I'll come and get them. And you'd be doing me the favor, at that. But it may be a couple of days."

We kids knew Dad wouldn't tackle wild bees in weather like that, hot and thundery, with heat lightning flickering in the sky at night. That's when even tame bees turn irritable and want to sting everything that moves.

Luckily the heat wave broke the next afternoon with a big thunderstorm. Afterward it was raw, chilly, almost like autumn—not good flying weather for bees. By dusk several of us kids were piled in the back of the old truck, along with the bee gear. The only two not going along were big sister (there was a square dance that night) and baby brother who, at the age of four, was too young to be out after dark.

Our new neighbor was surprised to see us. "You're not going to try doing the job tonight? By lantern light?"

"Sure," said Dad. "We'll catch all the bees at home, that way. No use leaving stragglers behind to hang around that flower garden till frost comes."

Well, the man rode along with us down an old logging road to show where the bee tree was. Once he'd pointed it out he retreated into a thicket to watch the show in safety.

A pale moon came out from behind the clouds, bathing the woods in ghostly radiance.

The hollow tree stood in a clearing. The small hole the bees used was high up, but there was a wide cleft in the trunk down near the ground, which showed an empty black interior like a chimney.

Using a can stuffed with cotton rags, Dad pumped smoke into that cleft. Inside the tree the bees set up an excited humming, and soon little curls of smoke came drifting out the bee hole. Satisfied that the cavity reached up that far, Dad pumped more smoke into the cleft until he judged the bees inside were stupified. Then with an axe he quickly notched the tree so it would fall just right.

Next came the turn of my two eldest brothers. Carrying up the long crosscut saw—one husky lad on each end—they set to work sawing down the tree. Since the trunk was hollow, within minutes it was nearly cut through and began to tip.

Instantly the boys dropped the saw and bolted for the creek, slapping wildly in the air and yelling "Run! They're coming after us. Ow, ouch!"

Now this was just play-acting to impress the watching stranger. In our part-Indian family, bee stings were rated little worse than mosquito bites, not worth fussing over. But incredibly, the boys stampeded that poor man and he ran with them all the way to the creek and even dived in.

When my brothers saw that, they waded into the water and splashed a bit too, so he wouldn't be the only one who was wet.

The tree had fallen as planned with the bee hole uppermost, splitting open for several feet and exposing the whole hive. The chisel, wedges, and crowbar we'd brought weren't even needed. At the impact, bees sprayed upward in the moonlight like a fountain, buzzing in alarm.

We waited until the bees settled down and were just crawling over the broken honeycombs.

Dad beckoned, and I stepped up. Although not as strong or fast as my elder brothers, I was able to stand stock-still and had the steadiest hand with a lantern. Not that it was the old lantern I held that night, though. The neighbor had lent us his big flashlight that threw a beam so bright I could have read a newspaper with it.

That was fine, for now clouds covered the moon. I stood close beside Dad. Working methodically and bare-handed in the white circle of light from the flash, he inspected the hive contents. Once he had located the queen and her guard, things went quickly. Queen, brood combs, and those with unsightly pollen cells went into the wooden hive box, along with a plentiful supply of honeycomb to last the bees through the coming winter.

Flowing into the box after the queen came a living blanket, the large hive population, now gorged with honey from the broken combs and therefore more docile.

Surplus honeycombs, shaken free of stray bees, were fairly divided by Dad between two dishpans and covered with oilcloth. There was plenty of that golden surplus too. With the hollow tree to expand in, the hive had been growing larger for years, instead of sending out swarms. Dad said later it was the biggest wild hive, with the largest honey stores, that he'd ever opened.

There was another wait, which seemed endless, while bees settled ever deeper into the box with its slightly askew lid. Then Dad brushed the last few slowpokes from the rim into the box with his handkerchief and closed the lid.

Our cautious neighbor was waiting for us back at the truck. He was so glad to see the last of those bees—and to gain a dishpan of honey without a single sting—that he didn't even seem to mind his wet clothes. I kept hoping he'd forget and leave the flashlight behind, but he didn't.

When we got home the whole house smelled of freshly baked bread. If there was anything better than warm bread, home churned butter, and plenty of honey, we didn't know it. The family celebrated the successful raid on the bee tree with a real feast in the old farm kitchen that night.

I've spotted a wild bee tree in my own woods now. It's just a small hive, not worth raiding, but I like to sit and watch the bees streak through the dappled sunlight on a summer day. It reminds me of old times, especially the night we raided the big bee tree, and I got to hold that wonderful flashlight for Dad. Also, how good the honey tasted smeared on a chunk of fresh bread.

For pleasures that rich, who would mind a few bee stings?

SEPTEMBER

25

THE SEPTEMBER LULL

There is a kind of lull between seasons, a few placid days when Nature itself seems to be taking a breather before changing directions. At such times a vague aimlessness takes possession of living things.

Earlier in the year it is called spring fever, but what is it in autumn? Birds no longer sing from their nesting areas, but it is too early for the great chattering flocks to gather for migration. Trees make no new growth, although the first light frost that will cause their leaves to color and drop may be weeks away. Nights are cool, while days often are as warm as midsummer.

We sense big changes coming soon—but not yet.

In the meantime the world seems drowsy, given over to insects. Clouds of gnats dance in the air. A legion of cicadas drone on in endless monotony. Crickets chirp. (One cricket has already found its winter quarters under the desk where I sit coping with that vague restless feeling that has no name.)

A curiously-branched brown twig is somehow stuck to the screen of my window. I glance away, but out of the corner of my eye I see movement. Yes, it did move. It is alive, a walkingstick looking in at me. Why do we see these insects only for two or three days in early fall, usually sitting on windows looking in?

Tired of squinting around this knobby fellow, I go

outside. More bugs. Yellowjackets and a dignified daddy-longlegs are having a banquet on the outdoor cat's plate, which was not quite licked clean. A pair of monarch butterflies circle a late-blooming milkweed by the fence.

Winter is coming, and there is work to be done. Still, the vegetables remaining in my garden may benefit from another couple of weeks of sun. Chopping wood? No, exertion like that is better kept for a cool day. And only an idiot would put in the storm windows in such warm weather. I take a leisurely stroll about the yard, instead.

A large furry spider, handsomely marked in yellow and black, has spun a web between a hollyhock and the corner of the house. This particular species of spider, too, I see mainly in autumn. A small crosshatched band like a crazy ladder, near the center of the web, marks its weaver as one of the *Argiope* clan.

Today she has made a capture.

A tiny green tree frog, its body about the size of a quarter, is evidently stuck in the web, making it bounce up and down like a trampoline. From the web's upper edge the spider gazes down as if uncertain where to bite such a fat, juicy, soft-skinned creature.

Taking up a little stick, I lift out the trapped frog. Its startled croak is more indignant than thankful, as it hops off into the grass.

A few minutes later the frog is back on his trampoline, bouncing it up and down. Aghast at such stupidity, I reach for the stick again. Then I pause. Don't tree frogs eat bugs too? Maybe the contest is equal, and I ought not to interfere.

The phone rings. I hurry into the house to find out who has lost a cow or rolled over a tractor. When I get back the spiderweb is empty. Both adversaries are gone, one of them forever.

Chores are waiting, but somehow my feet carry me down the trail toward the waterhole. Ahead of me on the path is one of the wood's prettiest creatures, a velvet ant. Its coloring is vivid apricot and onyx. Being of good size, nearly an inch long, it covers the ground swiftly. This one is wingless, therefore a female. It keeps to the path except for brief detours to peer sharply, feelers waving, down any small holes in the ground.

Known locally for some reason as the "cowkiller," the velvet ant preys on certain wasps that make their hives in old mouse holes underground, laying her own eggs there so that on hatching, the ant larvae eat up those of the wasps.

The velvet ant I was following soon found the kind of hole she was looking for and vanished into its darkness.

When I got back to the house I found that the walkingstick had moved a distance of eight feet, from the window to the screen door. Again it was staring into the house with keen interest, no doubt hoping to get inside and join my cricket in its haven.

No such luck, Mr. Walkingstick. I mistrust the cut of your jib. Scientists say you eat only tree leaves. If so, why is it, with all those bushels of green leaves still waving up there, that you aren't as plump as a caterpillar? Why are you built on such predatory lines, slim as a dragonfly? Why sit so motionless now, poised like a hungry praying mantis, even going so far as to disguise yourself as a twig, if not to lure some juicy beetle within reach of your needlelike head?

I ease myself into the house, taking care that the would-be invader stays outside. The cricket sounds a cheery welcome. By now the day is quite hot, so I head for the kitchen to make iced tea.

Just outside the kitchen window, under the eaves, hangs a large balloon-like wasp nest I have been watching all summer. Now a worker with a forlorn air is crawling about the paper walls, which are getting rather tattered, as if wondering where on earth to begin the needed patchwork, or if indeed there will be time before the frost. . . .

On the window ledge below, several other wasps have given up on a bad job and are dozing in the sun, enjoying the last few days of golden warmth allotted to them.

Those loafing insects have the right idea, I decide. There won't be many more such glorious days coming. Taking up the jug of iced tea, I retire to my lawnchair under the old whiteoak tree for a siesta. When Nature takes a breather, its wiser children follow suit.

26

THE DEN TREE

On that memorable day in autumn when I lost my heart to a bit of the Ozarks called Flat Rock Hollow, I had been climbing along the ridge on the old wagon road that marks the north boundary of the property. The real estate man at my side waved one hand airily at the dense greenery below and said, "What you want to do here is *park it.* I mean, clear out the underbrush. Cut down any trees that are spindly, lopsided, or crowded close together, so the grass can get enough sunlight to grow. In fact, you might just leave a tall tree here and there for shade—like in a park—and get rid of eyesores like *that*, right away."

That was a big dead oak, its remaining limbs skeletal against the azure sky. The lower trunk was shaggy with sheets of loose bark, while the upper trunk gleamed bare and white, pockmarked with small round holes among the fluttering scarlet leaves of Virginia creeper.

"Oak makes good firewood," the real estate agent said encouragingly.

I nodded. But already I knew the dead oak wasn't going to be chopped down, at least not in the near future. For I had spotted a great ramshackle nest high in the crown of the tree. A nest that size might harbor a family of storks, flamingos, or—more likely in this habitat—a pair of eagles!

After I came to live in Flat Rock Hollow I kept a close eye on that old dead oak. And the big sprawling nest did not belong to eagles or any other birds. It was—you guessed it—a squirrel's nest.

The animals were blackish gray with white to buff markings on ears and chins, and bushy tails, probably representing *Sciurus carolinensis*, the eastern gray squirrel. It was impossible to tell how many actually lived in the nest, because the tree was hollow. One squirrel would scold me, dive into the leafy nest, then shoot his head out of a hole under a broken limb several feet away an instant later, to chatter some more.

Or was that a second squirrel, making his presence known? I have never had more than four of the nimble little animals in view at one time, but so busily do they dart about the treetop, leaping into the nest and emerging at various popholes, that there might be a dozen of them.

Squirrels are most active in early morning, but in late afternoon at least one sentinel will be lying flat on a limb, legs dangling, basking in the sun.

Seasonally the squirrels desert the old tree for a time,

perhaps to harvest goodies in some other part of the woods, and the leaf nest falls into disarray. But they always return to rebuild the nest and assert their dominance vocally from the top of the tree.

Could I have cut down the old eyesore of a tree in their absence? Maybe, but that seemed a dirty trick. Further, I soon learned that squirrels were not the only inhabitants of this many chambered hollow oak, which proved a veritable apartment house in the woods. Woodpeckers lived in some holes, while various other birds had taken over and enlarged old cavities the woodpeckers abandoned. There was always a great deal of activity. Three distinct kinds of woodpeckers from small black-and-white-checked birds to giant redheaded ones drummed their beaks on the tree, searching for insects and grubs.

Brown lizards, and occasionally a metallic blue skink, would scoot about the trunk, probably after insects also. A raccoon mother raised two bright-eyed youngsters in a lofty den on the south side. A steady stream of wild bees kept coming and going to a small hole about twenty feet from the ground on the east side. No doubt this led to combs laden with honey.

On the ground level, a hole nearly a foot in diameter under a humped root served as an entrance to the subterranean den of some larger animal, perhaps a fox or woodchuck.

In the summer, small brown mouse-eared bats roosted under the sheets of loose bark during the day and prowled at night. A lone screech owl that kept the same hours, was most active in the fall and early winter. The snug home it used for a year was in a cavity nearer the top of the tree.

In the autumn, also, migrating birds liked to rest a while on the bare limbs that provided sunny terraces of the oak's friendly hermitage. Once a lordly goshawk used a high branch stub for a vantage point, surveying the world below through white barred eyelids until it spotted a reckless rodent and pounced on it.

The most exotic birds I ever spotted in the old oak seemed at least five times the size of a hawk. The sun coming out from gray clouds after a shower lit the woods, which sparkled with water droplets caught on twigs and orb spiderwebs as I made my way along the old wagon road.

Suddenly, from up ahead, a hideous squalling rent the air. I was not unduly alarmed, for I recognized that harsh voice.

Sure enough, there were a neighbor's runaway peacocks perched on a lower limb of the oak, looking as impressive as the storks or flamingos I had once associated with the huge nest. The pair sat close together, grooming their splendid plumage with their beaks, and the male did not need to spread his rainbow tail for the world to see he was a handsome fellow.

Seated on a stump, I enjoyed the colorful spectacle of peacocks courting for a while, before leaving regretfully to get on the party line and inform the neighborhood that the lost peacocks were alive and well in Flat Rock Hollow, thus ending their idyllic adventure in the wilds.

A charming story I was told later about this affair was that my neighbor had originally bought two peacocks, a male and a female, from the farm where they were raised, which was several miles away. The hen had died. So the cock had run away, back to the farm, and somehow lured another hen into eloping with him. Then either he was unable to remember the way back to his new home, or had reverted to the wild, for he and his new mate had been living in the woods for some weeks.

It was nice to learn that the enterprising bird's owner purchased the hen too, in order to keep the pair together.

The old oak has lost most of its limbs now. Soon, in a storm or high wind, it will topple.

But I hope that when it does fall, another dead tree nearby will be at the right stage, teeming with insects, striped with soft decaying wood for the excavating of cavities, to receive the present oak's high-living population as refugees.

Even after this venerable fallen giant lies prostrate on the forest floor, though, it will not feel the bite of an axe rendering it into kindling. Because as a log it will still serve as a home for many furry woodland creatures. Rabbits, foxes, skunks, and opossums will move into its cozy rooms. A small brown bear (I should be so lucky) may choose to hibernate there. Badgers, coyotes, ground squirrels, and woodchucks may dig their burrows under the log and feel more secure with a roof over their front doors.

So with a change in clientele, the oak apartment house in the wildwood may still be in business for years to come.

27

THE SAGA OF SCRAPPY THE RACCOON

Whenever the talk turns to wild pets at the home of Matt and Janet, I hear another episode in the saga of Scrappy, the fearless raccoon.

When it all began the family was still living in the city. The plan for moving back to a farm in the Ozarks was being talked of though. And the boy was already living that dream. He wanted a furry wild pet, preferably a raccoon, to raise. And he wanted it immediately.

So a raccoon was ordered—from a company on the West Coast that dealt in exotic pets. Soon the tiny orphaned animal arrived in a sturdy wood and wire box at a busy post office on a sweltering autumn day. The post office manager, an animal lover, tried to remove the baby raccoon from the stifling box to more airy quarters. But the raccoon, who evidently had begun to think of the box as his den during the long trip, put up a spirited defense, bloodying the hands of any who tried to invade his privacy.

When Matt arrived with collar and leash, he too got his hands nipped. So the raccoon, who that day earned the name "Scrappy," rode to his new home still inside the box.

For about a week he stayed in the box all day, coming out each evening to eat, drink, and cautiously explore the house. At last he chose the basement as his particular domain. The box-den, cleaned and spread with old towels, was carried there. It remained his favorite sleeping place, but he spent a great deal of time climbing too. He loved to sprawl on a rafter in the basement, legs and long, barred tail dangling, while he dozed with half-closed eyes.

With the boy, Scrappy proved an affectionate, eager playmate. He adjusted quickly to the leash and loved to be taken outside for walks, during which he was the center of attention among neighborhood kids. With farm dogs he might have had trouble, but evidently the city dogs and cats respected Scrappy, and there were no unpleasant incidents. The young raccoon was even tolerant toward romping puppies.

He was, however, keenly jealous of the boy's other pet, a turtle that also lived in the basement. Eventually the turtle disappeared and could not be found. Some weeks later, its empty shell was discovered atop a rafter, clean as a whistle.

After that the basement was in Scrappy's secure possession. He welcomed the boy and his friends there, also Janet who fed him. But he made every effort to drive away Matt, who occasionally went downstairs for tools. At such times a defiant Scrappy would station himself on the stairs, snarling and showing his teeth in a hostile display that said clearly, "No rivals permitted here; I'm master of this territory."

Except for thus staking out his territory, Scrappy was a well-behaved pet. He was not destructive of household furnishings. Raccoons, with their clever hands, have been known even to screw lids off jars, but Scrappy never tried to steal food. He would take morsels from Janet's hand very gently, so as not to nip her fingers. He liked meat and fish, of course, but also most fruits and some vegetables. Carrots, celery, and potatoes were acceptable, but he was not fond of sweet peppers. Apples were a great favorite of his, as was corn on the cob .

When he was given food, he would carry it to his water bowl and carefully wash it before eating. The difficult project of submerging a long ear of corn in a round water bowl, as the splashy ritual was performed by Scrappy, was very amusing for the family to watch.

As regards cleanliness, the raccoon was a model pet. He always washed his face with his paws, cat fashion, after eating, and kept his fur shining. He did not bury gnawed bones or old food scraps under the towels of his box, and he never soiled there. When that particular urge hit him, he set up a brisk rapping on the door with his paws until someone took him outside on his leash.

Janet says that often when she was sitting over her coffee in the kitchen, Scrappy would climb onto the table to sit in front of her. He would stroke her face with his warm little paws, while chattering away in a soft crooning voice. Janet submitted to this "loving" with some inner qualms, remembering Matt's bloody hands that first day, but the little fellow never harmed her or the boy in any way. His hostility was reserved for grown men who made abrupt movements.

Inevitably as Scrappy approached his full growth,

insistent urges stirred within him. He took to running off and staying out all night. His jaunts took him to Forest Park, which was just a few blocks from the house. Rumors spread that there was a very fierce wild raccoon loose in the park. The family confined Scrappy more closely and kept an eye on him—but to no avail.

The last time he escaped, it was by smashing his way through a glass window. Apparently he heard a summons to a fuller, more mature life.

This time the boy did not get his beloved pet back. So persistently did Scrappy hang around the raccoon enclosure in the park that in the end, a keeper yielded and put him inside with his own kind.

Later that year the family moved to the Ozarks. The farm has a large lake where Scrappy could have enjoyed searching for his own turtles and fish. It is surrounded by woods alive with sounds and smells, where he might be hunting right now under the goblin moon of autumn.

Well, maybe the Ozarks didn't need another raccoon. And at least Scrappy should have a longer life, safe from hunters' guns and hounds baying for his blood. He won't know the hazards of a hard winter, long droughts, or hurtling vehicles that destroy so much wildlife on the highways.

But somehow I feel that spunky little Scrappy, even if he had known of the dangers of life in the wild, at the time that his life forced him into a decision, would have chosen to live in a real forest instead of a park.

OCTOBER

28

BLACK WALNUT TIME

When colored leaves carpet the ground, and flocks of migrating wild fowl wing their way across chilly skies, it's time to take to the countryside with a burlap sack and look for a black walnut tree.

Back on Dad's farm the yearly nut gathering expeditions were high-spirited affairs. Even when the burlap sacks and baskets were filled to overflow, we kids hated to quit. My oldest brother always climbed the tree to shake the topmost branches vigorously, hoping to bring down some extra large nuts he had spotted up there. (He often did—right on his head.)

My own method was less dramatic. I walked in large circles away from the tree trunk, scuffling my feet in the dry leaves, on the theory that the biggest nuts roll farthest from the tree. This always turned up some giant ones. At least, I seem to remember things that way.

Our native black walnut is a beautiful tree of varied uses. Its majestic shape enhances the landscape. Along country roads its leafy boughs may shade travelers eating picnic lunches. Its wood provides many homes with beautiful furniture. Its nutmeats add a unique flavor to confections and desserts.

At the old homestead my parents had several pieces of walnut furniture, including a fine buffet that relatives from the city were always offering to buy for twenty dollars. But that was not the focus of my childish interest. Nor did I care much for the walnut candies, cookies, and cakes my mother produced which were raved over by adult visitors. In fact, I wasn't sure what the raving was about, then. At that time, black walnut flavoring was a bit too strong to suit my tastes. My favorite was the beechnut—mild, sweet, and easy for a child to shell with a flick of the thumbnail.

Shelling black walnuts was a chore—a job reserved for winter afternoons when the weather was too severe for outdoor activities. In some ways it was fun for three or four of us youngsters to work on those nuts. Usually we sat in the warmth of the kitchen around a big oak table, with an oil lamp shedding its glow if the day was dark. The frustrating part was that even working steadily—big brother cracking the nuts (and sometimes his thumb) with the hammer while we younger kids dug out the cracked nutmeat with small picks—a quart bowl would be not quite full when we had emptied the basket of nuts. And that was a lesser chore than the earlier hulling of the nuts, which left our fingers so stained it took weeks of washing with strong, homemade yellow soap to efface the traces.

A great-aunt of ours kept up the colonial art of extracting dyes from the hulls. Walnut yields not only many shades of brown from chocolate to beige (depending on the method used, reagent, length of time immersed, etc.) but also produced other colors as well—dusky gold, ochre, smoky blue, gray, and olive drab. Few people realize now that during the Civil War, the uniforms of both sides, the Blue and the Gray, were dyed with walnut stain. My own favorite was a russet of rich reddish tint. With all my heart I craved a sweater just that color.

On my tenth birthday I got the sweater from my great-aunt—a pullover with a V neck, hand dyed, hand knit, a thing of beauty. But alas, my birthday is in late spring, so that whole summer—the longest I can remember—the gift lay wrapped in tissue paper in Mother's cedar chest.

At last school days came again. Proudly I wore my walnut-dyed sweater all winter, on every occasion, always feeling smartly dressed and the envy of my classmates. That's when I was ten. Then all at once I was eleven, and the beloved sweater was taken from the cedar chest again. Had it shrunk? No, Mother was too careful for that when washing woolens. Yet the sweater was uncomfortably tight, and my wrists stuck out inches beyond the cuffs. I had grown.

Hiding my disappointment I handed over the russet sweater to a younger family member, one who unfortunately was a roughneck. In no time the elbows were poked out, patched, and then the back hopelessly snagged during a hasty crawl under a barbed wire fence.

Like the russet sweater, the walnut furniture of the old homestead is gone from my life, distributed years ago to older brothers and sisters who were married and had houses to keep things in at a time when I was beginning my wanderings with a dog-eared notebook and a sleeping bag. Still, I have many warm memories of black walnut time.

Recently at a community dinner here in the Ozarks, one of the families brought a black walnut cake. For old times' sake I took a piece. To my surprise it was the most delicious thing I'd tasted in years. The adult palate does indeed differ from that of a child, and now I know what all the raving was about. I went back for seconds on that cake, then tracked down the white-haired, pink-cheeked farmwife who baked it and coaxed the recipe from her.

It happens that beside a gravel road in my neighborhood there is a big black walnut tree straddling the fence line, its boughs hanging out over the road. A bushel of nuts could be picked up there in no time, already hulled by passing cars. To think that I've been driving over these flavorful nuggets for years, without knowing what I've been missing!

This year I'm taking along a burlap sack in my car.

29

A VISIT TO AN OLD CIDER MILL

One fine October day my niece invited me on a trip to a picturesque old mill for cider and doughnuts. It was keeping up an old family tradition, she said.

Her husband explained that the tradition had started four years ago with a simple drive to the country to see the coloring leaves. Then the kids had spotted some choice pumpkins at a roadside stand and had wanted some for jack-o'-lanterns for their Halloween party. Next they had discovered the old cider mill, which quickly became the center of this annual outing.

I was delighted to go along. I'm keen on family traditions—even new ones.

If the woods were browning and past their prime that day, nobody minded. Here and there a golden poplar or scarlet sumac retained enough leaves for a spot of color. The sky was intensely blue. The autumn landscape was alternately sunwashed or dappled with shadows from fleecy clouds sailing overhead. Here and there in the checkerboard fields, a few great rolls of hay stood like sentinels.

The mill was authentically old, as promised. It nestled in a pleasant valley. The water wheel (restored) turned slowly, cascading sparkling drops. In a placid millpond, ducks paddled among fallen leaves. A weeping willow on the far bank trailed bare branches in the water. Evidently the doughnuts were produced right at the mill, for the air there smelled good enough to make me wish I lived next door. Doughnuts and cider were made for each other!

After the repast, the kids guided me on a tour of the mill. The shaft of the waterwheel actually didn't connect to anything inside. The interior of the mill was modern and electric, combining the best of both worlds. We gazed in awe at great stainless steel vats, conveyor belts, and rows of gleaming bottles being filled with the amber nectar (filtered, pasteurized) of select apples.

After buying a supply to take home, we drove off, keeping an eye out for roadside stands offering pumpkins.

"I'll bet this really takes you back," said my niece. "Just like the old times, right?"

"Not quite." I gazed out the car window at the rolling countryside. "I miss the rounded haystacks that farmers used to make when I was a kid. They were hard to climb, but great fun to slide down, and just the kind of slippery place needed for playing king-of-the-mountain. And instead of being displayed at roadside stands, our pumpkins were piled in heaps where they grew in the cornfield. Oh, we had cornshocks in the fields too. We could burrow inside one in a game of hide-and-seek and never be found."

"What I meant," my niece broke in on these nostalgic topics, "was the old cider mill. Wasn't it like the ones your parents took you to, back then?"

"We didn't have to go anywhere. We had our own." I hastened to add that it was not a mill, only a two-man cider press worked by my oldest brother and Dad. During cider-making the press, bolted to a sledge made of planks, was moved to just outside the big double doors of the barn.

A circle of women—Mother, big sister, and various visiting relatives—sat with basins of bruised apples on their laps, cutting out any bad spots and worm holes. Only bruised apples were used for cider. Big sister sat nearest the press, and as the wooden cask beneath grew full she would call out for another one to replace it.

A dipper hung on a nail in a nearby tree for those with a thirst to help themselves.

For us smaller kids, there were mostly carrying jobs. Full baskets of apples to be carried to the women, trimmed apples to the press, buckets of waste to be toted to the pigpen or chickenyard.

The ground was always littered with our spillage before and long after the job was finished. Then the birds—especially robins and jays—would settle in for their feasting.

Sorting apples was another recurring job. A wagon piled high with them stood by the barn. We younger kids sorted these into baskets. The "keepers" were set aside to be stored in the coolness of the cellar. With a good harvest these would last until May. When the apple wagon was emptied, the team was hitched up and we would ride back to the orchard for another load.

It's a wonder all the apples weren't bruised, for they were all windfalls.

People used to the low, clipped-down orchards of today would have been amazed at our giant apple trees. Only a lumberjack could have climbed one, for all the lower limbs were pruned off the trunk so the mowing machine could be run beneath them.

If Mother wanted perfect apples, to give as a gift or to sell, my long-legged brother nicknamed "Lumberjack" would shinny up and toss apples to us on the ground until we all had sore wrists from catching. But it was great fun.

After cider-making came another fragrant job, cooking the apple butter in a big iron kettle in the yard. The butter had to be stirred constantly with a wooden paddle so it wouldn't stick to the kettle bottom and scorch.

Some of the cider had to be added back to the apple butter. Gallons more was sold to townsfolk or consumed at harvest partying. But much of it would be allowed to ferment on into cider vinegar. On our farm the apple butter would run out in January, the whole apples by springtime, but the vinegar had to last all year. As a child I felt this last was a waste of good apples, but now I can't see life without vinegar.

When I finished telling of the sounds, scents, and hectic activity of that bygone time, my niece's husband commented only, "It sounds like a lot of hard work." (This made me suspect that he had lived on a farm.)

"But fun," said my niece wistfully, although she had found nothing usable in my story to add to her growing list of family traditions.

Some days later, though, I scored big.

A bottle of cider had been pushed to the back of the refrigerator and forgotten. How it could ferment, being pasteurized, I don't know. But it was really perking.

"Spoiled!" wailed my niece.

"Nonsense," I grinned. "This is the real thing." I hadn't had the heart to tell her before that what she was calling "cider" tasted more like fresh apple juice to me. "At this stage, my mother made us hot mulled cider. It's served in mugs, you know. Your great-great-grandpa brought the recipe from England when Stonewall Jackson was a boy."

I showed her the ritual, the right amount of sugar and spices to add, and so forth. And everybody who tasted it, even the kids, loved it.

So hot mulled cider served in mugs, a "really old" family tradition, was added to the yearly Halloween party. Along with a contest and prizes for carving the best jack-o'-lantern face as well as the best costume. For next year, there is talk of adding a hayride too.

Now that's getting a bit strenuous for an old-timer. I don't keep such late hours anymore. And wagon rides, as I recall, are bumpy jostling affairs.

No, my idea of a perfect autumn tradition is being whisked along in a smooth running car to an old cider mill where the big wheel turns lazily in the sunlight, where ducks paddle in the millpond, and the air smells so good I wish I lived next door.

30

IN PRAISE OF PERSIMMONS

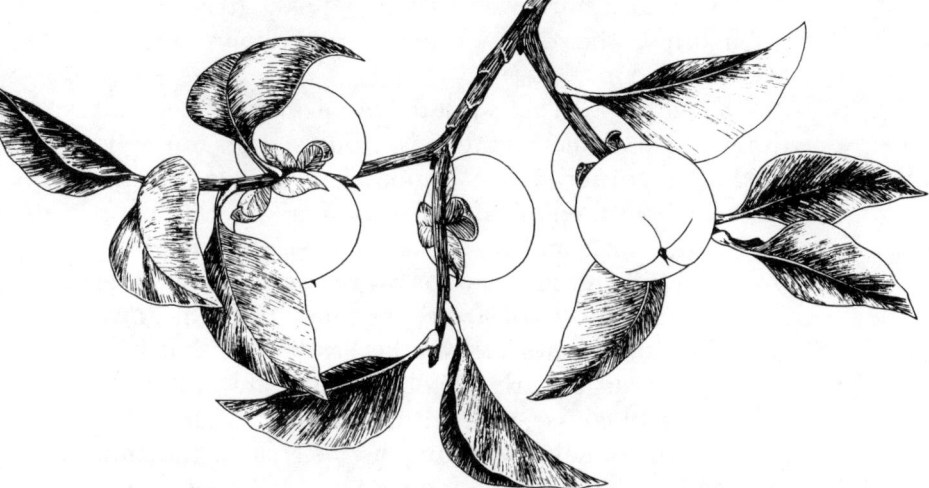

Does not the persimmon smack of the wild-wood? How little of the tamed orchard or trim garden in its sugary pulp! The town and all that means is for the moment forgotten, and you are in touch with Nature while you eat.

—Dr. Charles C. Abbott

Persons familiar from childhood with what Euell Gibbons calls the "Sugar Plum Tree" are lucky. I missed out on that. My own introduction to the persimmon came as an adult, when I ventured South with a group visiting caves. On a country road we stopped at a small restaurant, the kind blessed with a local cook who comes in to make pies and pastry.

The taste sensation of the whole trip came in the form of a hot persimmon pudding so deliciously gooey and figgy that we went for second helpings, despite the huge catfish dinner we had just put away. The nuts in that ambrosial pudding, were unmistakably hickory, which stirred up happy memories. My interest in the persimmon dates from that feast.

A small tree of the fencerow and edge of the woods, the wild persimmon (*Diospyros virginiana*) has a slender bole with dark, furrowed bark and glossy green leaves. When ripe, the dusky orange fruit shows a faint lavendar bloom.

When it's ripe? That's just the problem. No other fruit draws such diverse reactions from writers on wild foods. The writer of one manual says guardedly, "the astringent fruit, edible after frost, is popular with opossums, raccoons, and foxes," which doesn't sound too inviting. But another author waxes enthusiastic over "soft sugar lumps of fruit" recalled from his own boyhood, and speaks of trees he has known that bear "delicious fruit in early October, before any frost."

There is, of course, great variation from tree to tree in the size and quality of fruit, as well as in the time required for ripening. Individual variations aside, though, experts say that frost has nothing to do with the ripening of persimmons. The tree is a tropical immigrant and ripens its fruit slowly, in its own good time. For some trees, the snows of midwinter may well blanket the ground before the fruit is ready for feasting.

And if the persimmons aren't *fully* ripe. . .well, the alternative is celebrated in Frank H. Sweet's famous jingle,

> Have you ever
> On your travels
> Through the queer, uncertain South,
> Had a 'simmon—
> Green persimmon—
> Make a sortie on your mouth?

Astringent seems too mild a word for such a sensation. Yes, all too often those firm golden globes temptingly dangling on the bough are green 'simmons, acrid and puckery, best avoided. Only when they are dusky orange, soft and squishy, ready to drop off the stem, are they fit to eat. Fit? By then, they're luscious.

This lesson I learned the hard way, my first autumn in the Ozarks. Beside the road I filled a large bag with wild persimmons, carefully picking only unblemished, pale gold, very firm ones that had to be tugged loose from their stems. When I got home I tasted one (arrrrgh!) and decided to spread them on the ground outside to sun-ripen.

Over the next few weeks the mound of fruit grew somewhat smaller, as wild animals culled out any fruit approaching edibility, at night, but never did I find a ripe persimmon. Even a frost or two effected no change in the taste of these perfect looking persimmons. The mound continued to dwindle slowly, until at last the snow covered it.

In the spring a crop of beautiful little shiny-leaved trees came up on the spot. These I spaded out and bestowed on a visiting friend who professed to love wild persimmons.

I soon learned what I should have done. The only way to harvest persimmons is to spread a plastic sheet on the ground under the tree, shake the trunk gently, and retire with your treasures. If the fruit is ready to drop off, it's ripe. Easy as pie!

After that I began to collect persimmon recipes. In colonial times persimmon bread was a great favorite, being rated "better than gingerbread," which was then high praise. This versatile fruit also yielded persimmon molasses and vinegar as well as assorted pies, tarts, and puddings. And, of course, the chewy, long-keeping fruit leather.

Beloved of children were "sugar plums," which are

easily made. Simply slice off the stem end of the persimmon, gently press out the seeds, and dry the fruits on trays in the sun or in a slow oven. When dry, pack loosely in jars, sprinkling a handful of sugar over each layer. The sugar is not needed for sweetening, but it keeps the persimmons from sticking together. (You can omit sugar if you're counting calories or, like me, feel you're getting a bonus when you dip a hand into the sugar plum dish and get two treats for one.)

Used in place of raisins, chopped persimmon "sugar plums" give an interesting taste-of-the-wild fillip to fruit bars and granola mixes.

Of all the fine persimmon pudding recipes I have tested, though, even my current favorite does not taste like the one I remember from that little roadside restaurant. Perhaps we can never exactly recapture a given taste sensation. However, almost any persimmon-hickory nut pudding is a gourmet dessert.

Nature poet William Cullen Bryant, who was fond of wild persimmons, felt that with selection and breeding the fruit might prove a valuable American crop. So far, this has not happened. And somehow, I'm glad. Often I have bought some improved variety of fruit—big, beautifully colored, juicy looking—only to find when I sank my teeth into it that there was little flavor. The lovely coloring comes as a result of exposure to some gas or chemical, and the ultrafirm consistency is such that the grower could have kicked the fruit all the way to the market without leaving a bruise.

So let's rejoice that this native wild fruit is still untampered with. (The tame persimmon offered by many nursery catalogues is most likely *D. kaki*, a species imported from Japan. It's delicious, but not quite the same.)

There is a magic in eating the ripe fruit fresh from the

tree. It makes me feel like a twelve-year-old or an Indian as I lick the sweet sticky juice from my fingers, while my ears are nipped by a chill wind which promises snow, and all the other trees are bare.

NOVEMBER

31

ATTRACTING WILDLIFE WITHOUT A FEEDER

Undeniably, the best way to attract wildlife to your place is to start a feeding station. If you already have one, by early November your feathered friends will begin putting in a daily appearance in your yard. Some, with a touch of proprietary pride, you may recognize from past winters. Others will be strangers, possibly even of a species new to your area. Each day will bring new pleasures.

For years, while living in the suburbs, I enjoyed having a feeder. I was not afraid to leave for a winter vacation, knowing that "my" birds and squirrels would merely throng more thickly at the feeder down the street, or the one on the next block—places they were already visiting sporadically to sample the fare.

Living alone in the country is different. Here my nearest neighbor is a half-mile away. The closest one with a bird feeder is over a mile off, as the crow flies—which is too far for many birds to commute on their daily feeding circuit. Now I can easily avoid taking a winter vacation, but there is no guarantee that I won't be hospitalized for pneumonia or a broken leg. So I have hesitated to start a feeder and risk disappointing, or even starving, my wild friends during some blizzard when they might need my help to survive.

Other folks in the same boat are those who, at present, use their Ozark places for weekends and vacations only—but still want the pleasure of watching wildlife while they are here.

In such cases, an alternative to the feeding station is to improve the habitat gradually, so wildlife will not become dependent on handouts but will be self-sufficient throughout the year.

November is a good month for taking stock. When the harvest is in and firewood cut, the weather may still offer sunny days when the ground is not frozen—ideal for landscaping jobs. With a bit of care you can choose permanent plantings that are not just beautiful to look at but also bear fruits or seeds that are choice food for birds.

Verne E. Davison lists a number of ornamental trees, shrubs, and vines that wildlife is known to feed on. Bluebirds, he says, feed on honeysuckle, cotoneaster, viburnum, mountain ash, and privet hedge. Cardinals patronize arborvitae, beautyberry, dogwood, and camphortree. The yellow-shafted flicker feeds on magnolia, holly, and spicebush, while the evening grosbeak prefers box elder, catalpa, and pine (eastern, white, loblolly, and Virginia varieties). Hummingbirds visit azaleas, mimosa, coralberry, Japanese honeysuckle, trumpet vine, and weigela, as do bees and some butterflies.

Wherever I live I will always have a fragrant mock orange bush. As a child I saw as many as a hundred tiger swallowtail butterflies, a joy to behold, hovering about one large mock orange by Grandma's garden gate.

Other plantings not only attractive to the eye but a boon to wildlife include chinaberry, Japanese barberry, snowberry, cherry elaeagnus, matrimony vine, Christmas mistletoe, nandina, pyracantha (firethorn), smoketree, and tulip tree.

The stately maple is desirable not only for its seeds

and buds (which are eaten by purple finch, grouse, turkey, pine siskins, and squirrels) but also as a preferred nest site of many birds.

The big advantage of such dual purpose landscaping is to bring the birds near the house, where they can be watched.

Water is a basic need of all life. Putting in a good pond will attract animals and birds in your vicinity throughout the year. And if you sow watercress along the banks, both you and the mallards can enjoy this succulent salad green. A packet of sunflower seeds sown along some fence can also be a good investment for bird watchers.

Very important for the year-round survival of wild life, of course, is leaving as many of your natural food sources, cover, and nesting sites as possible. When thinning the underbrush in wooded areas, try to spare the dogwood (a choice food of bluebird, robin, bobwhite, grosbeak, and mockingbird), hackberry (robin, Swainson's thrush, waxwing, turkey), persimmon (relished by mockingbird, catbird, robin, turkey, pileated woodpecker), black locust (eaten by red-eyed vireo, warbler, downy woodpecker, etc.), and Virginia creeper (favored by bluebirds).

The serviceberry, wild cherry, blackhaw, wild grape, smooth and staghorn sumac, wild plum, mulberry, and elderberry also provide valuable wild food.

Deer browse on buckbush and other undergrowth, and brushpiles offer shelter to smaller mammals.

Aldo Leopold's concept of "idle acres" is pertinent here. Most rural properties have a bit of land too rugged, or sandy, too dry or marshy, to do much with. Yet each is, in its own way, a valuable habitat for wildlife. Thus it is worth preserving. Also, a dead or diseased tree in the unusable land will offer rich insect food or dens. Perhaps a field adjacent to the woods may be allowed to revert to wild grasses and weeds which feed many kinds of birds over winter.

Leopold warns, "Keep cows, dogs, and mowing machines out of your idle acres" during nesting season, so that ground nesting birds like quail, grouse, pheasants, and turkeys can raise their broods safely in the tall grass.

You'll feel it's well worth it, when you see your own quail or turkey flock coming in to your pond to drink. Soon your main problem may be that your family becomes so attached to this particular flock that you will have to do your hunting farther afield.

November is also a good month to put up new bird houses. In Missouri, male birds may arrive to claim territories as early as February, and it is desirable to have the new houses weather a little before tenancy. Even if you use unpainted wood (as the birds prefer) weathering helps the house change the carpenter shop smell for that of the great outdoors.

And if the evergreens you planted aren't big enough yet to shelter birds during winter storms, you may want to put up roosts. What kills birds in a severe winter is not the cold per se, as their feathers insulate well enough ordinarily; but being chilled by an icy wind or getting wet can be fatal.

An easy roost to build is an oblong box with an overhanging roof above its one open side. Perches inside can be made of quarter- to half-inch doweling. This should be attached to the protected side of a building or in an ell.

Another style, which can be hung on trees in the open, looks like an outsize milk carton, with a round access hole near the bottom. The rods inside should be staggered, so that the birds are not sitting directly above one another.

166

Bluebirds and wrens are only two of the species that appreciate dry roosting boxes in bad weather.

By improving the habitat of your land in these and similar ways, you can build up an interesting wildlife population gradually. Then, when you see your way clear to starting a feeding station, you will be delighted to find a multitude of wild friends already on the scene, waiting for the gala opening day.

32

UNCLE JOHN'S WALKING STICK

Noted woodsman Bradford Angier advises carrying a pole when traveling over rough terrain. This can be used for probing ahead to insure safe footing, or supporting your weight in case of a sprained ankle. It is handy during wading to check the depth of the water, for balance, and for bracing yourself against a swift current.

When crossing a frozen river, a pole is useful in locating patches of rotten ice, or finding the depth of snow covering. If you do plunge through, the pole quickly turned sideways to catch on the edges of the ice can break your descent and help you to climb out of the water.

In Europe such poles or staffs have been held essential for foot travel since Robin Hood and Little John sparred lustily with theirs to see who should first cross a stream via a fallen log. Over the years the pole has evolved there into the popular walking stick which no European would be without on a hiking trip.

Some of these sticks are fancy indeed, perhaps weighted with lead like the "Penang lawyer" or hollowed to contain a sword, but they are always stoutly made. The Englishman may refer to his as "my trusty cudgel"; the Irishman calls his *shillelagh*; to those who live in the Alps it is an *alpenstock*. Now, as in Robin's time, it serves the dual purpose of self-defense and an aid to walking.

A friend of mine once saw a group of Fench nuns on a foot pilgrimage to a country shrine pass along a road, each with a formidable walking stick.

My Uncle John's stick, of seasoned ash, was in this grand tradition. Mind you, it was not a cane; it was not topped by any flimsy shepherd's crook but by a heavy knob shaped to fit the hand. This knob, weighted with lead, unscrewed to reveal a cavity hollowed in the shaft, that held a slug of bourbon to sustain him in case of accident or, as the proud owner said, until the Saint Bernard arrived with the brandy keg.

Uncle John carried his walking stick in the city as well as in the country. This was because he was slim (though wiry) and would not have measured five foot six in cowboy boots. And the city streets were not that safe even then.

My grandmother's elder brother, he was an old man when I was a small child, but his gait was springy and his curiosity about nature as keen as ever.

When visiting Dad's farm Uncle John went on long solitary walks in the woods, and always returned with some interesting story featuring his amazing walking stick.

On winter evenings when we were gathered around

the warmth of the wood stove, with the wind howling outside and sleet rattling the windowpanes of the old farmhouse, Uncle John would tell his best tales. How he had once scattered a pack of hungry wolves that were trailing him. How he hit a big black bear on the nose in a berry patch and made it flee. How he chased a wildcat from a cave to gain shelter on a cold night. How he laid out a rabid skunk and killed many a rattlesnake.

I believed all these stories then—and still do. There *were* some black bears, wolves, and bobtail cats in the woods when Uncle John was in his prime.

Besides, I'd seen him in action with his walking stick.

Once we went to visit a relative in town. A big dog that was the terror of postmen and passersby there leaped a gate and tore down the street, eager to taste Uncle John's thin shanks. Quick as a fencer the old man lunged with his stick, which slid neatly under the dog's collar beneath its chin.

Lifting up on the stick then, he forced the dog to rise and walked it on its hind legs up to the owner's door.

When the astonished housewife answered his rap, he politely raised his hat with his free hand and asked, "Is this your dog, Ma'am?"

The choking dog, abruptly released, dropped to all fours and slunk into the house, tail between its legs.

Uncle John was a hunter, too, although he never carried a gun. Often he came back from a walk in the woods with a fat rabbit, squirrel, or grouse for dinner. Well, that didn't perplex me any. He not only had his walking stick, but also was a deadly shot with a snowball, green apple, or stone.

Sometimes he brought in a sizable fish catch, and that did bother me. How could he catch fish with a stick or a stone? The force of the blow would be dissipated by the surface of the water.

Then one day I discovered the answer.

Several inches from the tip of the walking stick was a metal band, evidently for reinforcement. This band suddenly rotated in my hand, and the end of the stick dropped off to reveal a sharp metal point. It was a homemade job; the wooden stick had been sawed in half, the business end of an icepick inserted, and a hole drilled down in the shaft tip to sheathe it. Then the join had been hidden by adding the metal band.

Uncle John had simply been spearing fish, as the Indians did.

At first the old man was a bit irked that I had found out his secret. But soon he started showing us kids how useful the tip was for picking up a dropped handkerchief, or toasting marshmallows three at a time at a bonfire. Nudging me, then, he chuckled that there was one more secret to the walking stick—one that I wouldn't figure out so easily.

I never had a chance to try. The next winter Uncle John died in the city, and the stick was not among his things that were shipped home. To this day I don't know, and can't even guess, what that last secret was.

When I go for a long hike now, I always cut a stout oak or hickory sapling and trim it into a walking stick, remembering Uncle John and his stories. The stick has saved me from some nasty falls, even as Brad Angier foretold. But so far I have had no animal adventures with it, beyond rescuing a few drowning bees from a creek. Still, who knows? There are plenty of copperheads in this region . . . coyotes too.

And recently my neighbor saw a bobcat in a sycamore tree.

33

THE OZARK HOUND

When hunting season rolls around and nights are frosty, I think of a certain Ozark dog.

I never knew her real name, or that of her rightful owner, in the few delightful (for me) months of our acquaintance. Somehow it never seemed right to give her a whimsical name of my own invention; she was too direct and earnest about her mission in life for that. Now I think of her simply as the Ozark Hound.

A wise man once said, "Biology is destiny." That was particularly true of this dog. It was in her genes: she was born and bred for the great game, the hunt, and would be content with nothing less.

This remarkable canine entered my life on a foggy morning. As I was filling the food pan of my outdoor cat, something stirred in the tall grass. A small dog had been curled up, watching my door. Slowly it rose and limped forward, only a half-grown puppy, tail at a timid half-mast, then beginning to lift and wag as the animal became more sure of a welcome.

"It's my chow time too," the brown eyes seemed to say.

A feeding was long overdue, in fact. The lost, footsore pup was so thin I could count ribs. Her neck had been rubbed hairless, showing old gall marks from a collar or chain. A new, raw wound near the end of her tail bothered her, and even while eating she would stop, whimper, and lick the sore.

Now I wasn't ready for a dog yet. Besides, she belonged to someone else, as the collar mark showed. Nevertheless an hour later I was in town, being told by the admiring vet how lucky I was, as he gave shots, first aid, foot ointment, the works. He praised the neat build, the slim legs that could run for hours, the deep chest with ample lungs, the poised thoroughbred's head.

"And I'll bet she's got a good nose," the vet added. "This one won't lose the trail."

The little foundling appeared to be a purebred hound of ancient lineage, a prestigious breed of hunting dog much prized in the South. My neighbors were equally impressed when they saw the pup.

Failing to get a response to my newspaper ad and inquiries, I had myself a dog. One to take on my walks in the woods—always a pleasure. Surprisingly, though, she did not frisk about joyously on such occasions. In the woods she became deadly serious. Often she stopped stiff-legged to smell the breeze, or went into circles, then put her nose to the ground and scuttled off into the brush, her legs working like pistons.

I was supposed to follow her, but I didn't know that. Instead I would be stooping to examine a patch of cord moss starred with crimson peristomes, or leafing through a pocket guide to identify an unfamiliar bird.

In the distance I would hear my dog emitting yelps, or the indignant sounding "Rooo-rooo-rrooo...." But the few times I searched her out to see what the fuss was about, I never caught a sign of her quarry.

That the Ozark Hound wasn't really happy with my fuddy-duddy lifestyle became clear as months went by. Loftily she refused to learn the fetching of newspapers and slippers. When I bounced a rubber ball she smelled it once, then walked off with a disgusted air. Nor did she ever bark, although at night I would hear her bell-like yodeling off in the woods sometimes as she kept a lonesome (and vain) vigil by a den tree or interesting hole in the ground, waiting for me—or anyone—to bring the

But she never really barked, not even at strangers.

In fact, she seemed eager to be buddies with repairmen, delivery boys, and any stray hiker who dropped by. Her silliest and most fawning display, though, was put on for three grim-faced, unshaven men with rifles who stalked out of the woods, looking exactly like escapees from a prison. They were cattle owners after a coyote, but how could she have known that?

Thus, unsubtly, the Ozark Hound indicated that she was above mere watchdogging. Any common Jack-of-all-trades dog could bark at strangers. She was a professional and meant to stick with her specialty.

And I still think it miffed her not to have a real name of her own, as a dog should. Sad to say, when I spoke to her, it was simply "Puppy." Or even worse, "Puppy-wup." Adding insult to injury, even this poor excuse for a name was made a term of reproach.

"Anyone would know you're still a puppy," I scolded her when she tore sheets from the clothesline for flapping at her. Or, "Just a big dumb puppy-wup, you are! Don't even know when you've got it good," when she chewed up a pair of muddy boots that seemed to have been left outside purposely for her to play with.

No, I didn't understand her. You can't bribe or buy the affections of a real, down-to-earth, proper Ozark Hound. Choice food, the best vet's care, a swank dog-house with cedar chip floor to repel fleas, freedom from beatings or chains or wire pens—all this can't make amends for denying an honest animal its birthright, frustrating the destiny in its genes.

This truth finally came home to me one icy day during hunting season. I was stacking firewood when a pack of dogs broke into full cry and hunters' shouts echoed up on the ridge.

Like a shot the Ozark Hound dashed toward the excitement.

At the edge of the woods she paused an instant, one forefoot uplifted, and glanced back at me. Every muscle in that superb body, an ungainly puppy's no longer, was quivering with tension, ready for action. "Come along," her look seemed to say, "I'll show you real life!"

But there was a lot more firewood to be stacked before dark, so I kept at it. Moments later her soprano voice joined the others of the hunting pack, and dwindled in the distance.

She never came back. Maybe her former owner found her, or another hunter figured she was lost and adopted her. At least she chose the life she wanted.

I still don't have a dog. Maybe that's because the pup's clear judgment of me was too painful to my ego. One consolation is that none of my close friends could have met her high standards either.

Retired folk are a peculiar breed too. They may remember fondly a farm childhood, may truly love the country, but were altered and made more sedate by years of drudgery in the city. They tend to live in remodeled farmhouses or mobile homes with a fireplace, have tiny dogs named Cricket or Tiffany that stay in the house . . . and their idea of exercise is the daily walk up the lane to the mailbox.

When they do go out, they carry binoculars or a camera, not a gun.

No genuine native-born Ozark Hound will put up with that. Not for long.

DECEMBER

34

A WALK WITH CATS

One of my favorite stories as a child was that of Robinson Crusoe walking around his island kingdom with his cats. I could just see the younger cat scampering ahead, putting sea birds to flight or sniffing at the holes of fiddler crabs in the sands, while the older cat paced companionably at its master's side.

When I moved to my own wooded solitude I aimed to take a leaf from Robinson's book. Mine were city cats that had never set foot on grass or scratched on a tree. Taking them for walks, introducing them to the fascinating realm of the wilds, were things I looked forward to doing.

So on waking that first day in my new home, I picked my way around the clutter of unpacked boxes to lead the way outside and greet the rising sun. The cats dashed out eagerly.

"Come on, fellows," I called. "Taffy, Pete, kittens. Let's take a stroll down to the creek and investigate a real water hole. Learn about frogs."

I went down the path, drawing in deep breaths of clear mountain air with a faint tinge of wood smoke, reveling in the bird songs overhead.

After the first twenty paces, however, the cats were not with me. I went back, patted furry heads, spoke coaxingly. We started off again. But again there was lagging, with much studied licking of fur and a couple of piteous meows. In vivid cat language they said there was some mistake: they would get hopelessly lost; they were tired already, and footsore. When I stooped to pet a feline, it would instantly roll over on its back, legs in air, and vulnerable belly exposed—the ancient animal "I surrender" signal.

All this time the house behind us was clearly in sight, the path wide, and our destination ahead was teeming with interesting things like flying insects and barking squirrels. No sale, though.

Weeks passed, then months. My walks in the woods remained solitary. No cat ever felt like going along. They were always having an off day or needing a nap. Oddly enough, though, somehow they found ways to familiarize themselves with the woods behind my back, for I began to spot them during my own jaunts up on the ridge, or on a gravel road a mile from home.

At such times a cat would be acting exactly as if it knew its business there. Until it saw me, that is. Again would come the piteous meow, the sorepaw licking routine, the rolling bit, then a coy but steady retreat homeward with me following.

And it dawned on me.

The cats were not afraid of getting themselves lost at all. But they were afraid I would get lost, never find my way back to the house, never again fix tasty plates of food or open doors for them or build a warm fire in the stove on cold winter days. (Cats have no high opinion of me anyway, I suspect. How can any reasonable animal respect a being who can't even tell that a rank smelling fox has passed by not ten minutes ago, or that a quail is crouched just behind a bush?)

So despite the fact that most of the cats had explored the woods by now, none of them would go for a walk with me.

Until a windy December day I'll never forget. That morning I was chopping wood. After a while I became aware of steady cat traffic on the path to the ridge. Up and down, continuous coming and going, with an air of excitement. Often a cat about to enter the woods would look back to see if I was interested also. Could it be that they wanted to be followed?

I went. Sedate Chinook, patriarch of the cat clan, even waited for me and trotted at my side. I really felt like Crusoe when we overtook the kitten Smith sitting on a stump, and from then on the furry pair ran on ahead, neck and neck.

The way led over the ridge, under a barbed wire fence, across a gravel road, through a field, and into another woods. There my other cats, plus one from a nearby farm, were sitting in a semicircle watching a thicket of small cedars.

A kind of scuffling was going on in the thicket. Pointed cat ears pricked up intently. Something interesting, perhaps dangerous, was lurking in there. A big animal, by the sound of its threshing. Then a long desolate wail split the air and raised goosebumps on my skin.

What was it? The sound was one I realized I had been hearing on and off, all morning, above the wind's whistling. But it had come from a distance, was clearly not on my property, so I had paid scant attention to it.

The eerie cry came again. The cats turned their heads to look at me, questioningly. Of course I had to go in and investigate.

There was blood on the ground. A big animal, all right, with a badly swollen foreleg clamped in a steel trap. It was Junior, a dog I know, and he was of a size to knock down a grown man when in a playful mood. And Junior was usually friendly as a pup.

But today he was not answering to his name, or to "good doggie" either. Wild with pain, he snapped at my hand each time I reached for the trap.

Frustrated in attempts to free him, at last I headed for home to get help. Junior's owner did not answer the phone, though. Later I learned that he had been out driving the roads, searching the ditches for his lost pet. The landowner could not be reached either; he lived in the city. Nor could I guess to what local person (if any) he had granted permission for trapping.

Finally I phoned a neighborly farm couple.

"We'll be right over," they said instantly. "Hold the fort."

When I flagged them down on the gravel road, they climbed quickly from their truck, the man carrying a large cage, his wife with an old blanket over one arm. We proceeded through a suddenly catless landscape to the cedar thicket. The blanket was thrown over the dog, the trap unsnapped from the mangled leg, and the still firmly trussed Junior was deposited in the cage with a brisk yet gentle efficiency that astonished me.

Within minutes the farm couple were waving good-bye, on their way to the vet's.

Junior's leg bone was shattered, requiring a cast. But it turned out he was not in bad shape—not dehydrated yet. Thanks to the cats, he had been found in time. Not until some time after the cast removal, though, did Junior become his old frolicsome self.

As for the cats, since that windy December day not only have they refused to go on any hike with me, they actually begin crying and fussing now every time I start toward the woods. They have a new worry, I guess— that I'll step into some trap and never come back.

Well, that's one worry Crusoe's cats didn't have— traps cleverly covered with dead leaves. But come to think of it, those felines knew their master couldn't get himself lost either. Not on an island.

35

WARDING OFF CABIN FEVER

A thin dusting of snow covers the earth. Up ahead a small dark bird is doing the chickadee cha-cha. That's a hop forward, two quick scratches, a hop back, peck, then do it all again. I halt in midstep, not wanting to disturb its breakfast—but too late. The bird flies off.

Beside the trail is a patch of frost-curl plants, one of nature's mysteries to me. Nobody here seems to know the name of this weed. (It is wiry stemmed, much branched, about a foot tall, has opposite leaves, and bears tiny purple flowers in late summer.) In cold weather, even when the ground is bare, the base of its stalk will be sheathed with curls of ice that resemble designs left "when Jack Frost paints the windows," as the old saying goes. How does it happen, and why on just this one plant?

Now snowflakes start hitting my face, and soon the Currier and Ives landscape of black, brown, and silver is nearly obscured by a shifting white curtain. I hike on.

Why am I out in such weather? Well, actually I'm warding off cabin fever.

An underlying cause of cabin fever, known in its milder form as the winter blahs, is now held by experts to lie in our circadian rhythm. It seemes that, like plants, we humans need a certain number of daylight hours for our physical well-being. The effects of prolonged light deprivation during the short dark days of winter include a drop in body temperature, slower respiration and heartbeat, decreased blood supply to the brain and extremities. The body goes into lower gear, as if readying for slumber.

When at the same time our lives are demanding maximum physical and mental efforts from us, the result may be strain, lassitude, irritability, fatigue, headache, and a feeling of depression.

One way to prevent cabin fever is to maximize one's exposure to what little daylight there is. I try to get outside some each day, even if the weather is dreary. A surprising amount of solar wave-lengths, including ultraviolet rays, get through what seems to be a dense cloud cover. (A friend of mine who is an albino once got a bad sunburn while skiing on an overcast day.)

Winter sports are fine but often require weekends, companions, and travel. What is needed is daily, outdoor activity. Walking, for example. There are many reasons for walking in winter. One retired woman who bundles up and takes long walks says, "I'm not likely to run into any copperheads." Her husband, who walks and also jogs, feels that cold weather is more invigorating.

For myself, I like to observe nature at all times of the year. Being not too skillful at the "bent blade of grass and broken twig" school of tracking, in winter I find signs of wildlife activity more abundant and easier to spot, due to the snow and the soft earth of thaws. One game I play is to put these signs in different categories, to identify just which animal did what.

The first category is "marks left on the environment." This includes the higher browsing of deer on bushes and twigs, the lower browsing and bark gnawing of rabbits and other rodents. The remains of trees felled by the chisel teeth of beavers may be seen near streams. On the slopes above, trunks are scored by the rubbing of deer antlers. In mild weather a line of raised, buckled earth will indicate the passage of a burrowing animal below ground.

The next category is: "things left behind." A lone tail feather from a turkey, quail, or bluejay would be shed in passing, but a scattering of several small feathers from the breast or belly indicates a bird kill on the site, even if there are no bloodstains. And then there are droppings, a very specialized subject. I can recognize deer droppings, rabbit pellets, pigeon lime, and the dung of opossums if there are flat brown persimmon seeds in it, but that's my limit so far.

Another category is "imprints." Once, in the snow, I saw a perfect fanlike outline of a hawk's wingtip made as it overbalanced while grappling a mouse. A set of much scuffled pawprints followed and half obliterated, seemingly, by a pursuing snake, was made by a large opossum dragging its tail. A smaller, arrow-straight set of footprints that started in the middle of nowhere and ended abruptly near an oak tree was left by a squirrel's frantic dash. Overlapping hoofprints of varying sizes on a hillside showed where a family of deer went in single file toward the waterhole.

One bewildering set of large bird tracks proved, on closer study, to reveal where a flock of grouse emerged from a snowdrift in which they had taken shelter during a heavy storm.

I enjoy walks in all kinds of weather—sun, rain, fog, or snow—but avoid one thing. That's a high wind. Sylvan Hart, the last of the Men in Buckskins, once said, "A high wind will kill you. You die like a fool."

So on very windy days I stay indoors, with a lot of bright lights on. If plants will grow and bloom under incandescent and fluorescent lighting, some benefit may accrue to humans also. One bright light is over the stove where a pot of soup will be simmering. Another light that is always on when I am at home in winter is right above a table where there is usually a partly assembled jigsaw puzzle. (Preferably the scene will be sunny, with green trees overhanging a blue lake, and rolling meadows so lush I could almost walk out into one and pick a buttercup). Dunno about my circadian rhythm, but such things do wonders for my morale.

In this area, winds tend to die down toward evening, so it is often possible to take the daily walk after dark. The stars of winter skies seem more brilliant than those of other seasons, perhaps because the air is clearer, although with the leaves off the trees, more sky is visible too.

Appropriately, the most eye-catching constellation in winter is Orion the Hunter.

Sporting his sparkling three-star belt, the hunter strides across the night sky, followed closely by his dogs. The eye of the Big Dog is Sirius, the brightest of all stars, while the Little Dog has Procyon, also of high magnitude, in its tail. The constellation Lepus, the elusive rabbit they are chasing, is a bit harder to spot. But that's appropriate, too.

One of the names of an old Indian I once met was Drinker of Starlight, from his habit of hunting at night.

Starlight? Those millions of stars up there are really huge suns, all pouring out radiation which reaches us across vast gulfs. If even the pale reflected light of the moon can influence the growth of earth's plants and the spawning of sea creatures, who can say that these nameless stellar light-rays have no effect on us? Maybe they do.

In any case, drinking in a bit of winter starlight helps me beat off a bout of cabin fever, every time.

36

BEANS FOR CHRISTMAS

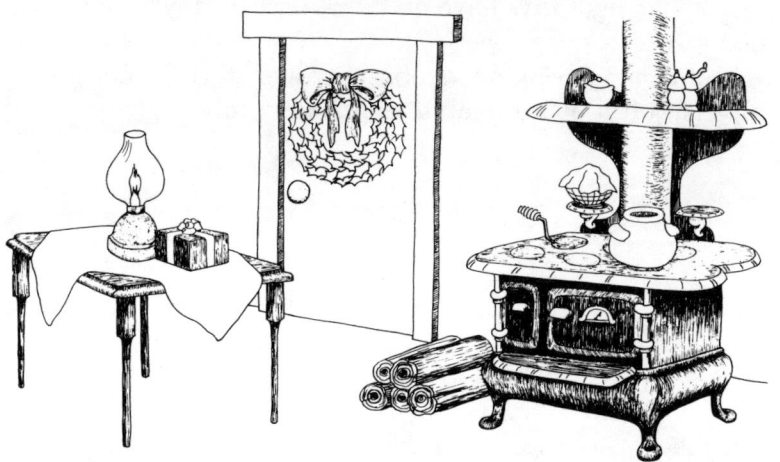

When old-timers meet on Christmas morning in Jake's Prairie they greet each other with a cordial, "You got the beans on cooking yet?"

Many families descended from early settlers here keep to the old custom of having beans on Christmas Day to bring good luck to the household for the coming year. No matter how sumptuous the holiday feast, they say, on your dining table you should always find room for that humble pot of beans cooked with a hambone or bacon rind—just for luck.

One of my favorite old-timers, white-haired Amos Barclay, held firmly to this custom. Well versed in Ozark folklore, he also believed that if you enter an empty room to find a rocking chair moving all by itself, it means there will be a death in the family. That on the second week of April the whippoorwills will return each year to sing in these hills and hollows. That when the leaves are so thick on the hickory trees you can't see blue sky through them, it means a cold winter is coming.

Such curious lore, as well as the canny woodcraft handed down from his grandfather, were part of the daily life of old Amos, who used his knowledge of the wilds to live comfortably, in a way that suited him, on a small retirement pension.

A keen hunter, he spent a great deal of time in the woods. Much as he enjoyed stalking game, though, he shot only for the table. Savory fried rabbit, squirrel stew with dumplings, roast opossum or raccoon stuffed with apples enriched his diet (while stretching his slender grocery allowance) as well as tender quail, turkey, and venison in season.

With fishing, too, Amos combined business with pleasure. At certain times of the year he went to the lake and fished almost every day. Panfried fish with greens and cornbread made one of his favorite meals.

And what about those doldrum stretches when the fish just didn't bite? Amos would find *something* to bring home and fix—maybe even gourmet items such as turtle soup, steamed crayfish, or batter-fried froglegs.

If the foraging really did prove fruitless, or the weather was too nasty to go out, then Amos might open a jar of venison to go with his potatoes and cornbread. That's right, a jar—not a tin can. For this oldster, a handy man in the kitchen, put up any surplus meat in quart jars so that he always had something available in the pantry. He canned fish, too.

Asked how he served his home-canned fish, he offered two recipes. One, make a creamed gravy with onions in the skillet, add the fish chunks, simmer fifteen minutes, and serve over cornbread. The other recipe was one which I had eaten in Maine. Flake up the fish with an equal amount of grated potato, make patties, and fry until brown on each side.

Both of these are good hearty dishes when the wind whistles around the house and icicles hang from the eaves.

In his walks in the woods, Amos did not neglect other goodies in nature's larder—mushrooms or a sassafras root for making tea. His homemade blackberry and gooseberry jams and wild plum jelly were delicious.

Amos was eighty-one when I first met him several years ago. But his gait, appearance, and energy level were those of a robust man in his early sixties. He attributed this vitality to his habit of eating beans at Christmas. But certainly his way of life had a good deal to do with it, also. He would be out of doors much of each day, hike several miles to visit a friend, plant a large garden each year, and cut his own firewood.

With only his two dogs for companionship, Amos lived in a small trailer hooked onto a shack with a corrugated tin roof. The trailer, which was his kitchen and living room, was wired for electricity, but the shack (which served as his bedroom and sitting room) was not. This primitive setup suited the old fellow's independent spirit.

When there was a severe ice storm and power blackout, Amos would merely open the door of his wood stove, enjoy the firelight flickering on his shabby, comfortable furniture. If he had not finished his chores when the power went off, he could light a hundred-year-old oil lamp, he would use a lantern when he went outside.

On nights when the temperature dropped below zero and people in modern houses worried about frozen water pipes, Amos relaxed. The long-handled red pump on his back porch brought up water from the well only when it was needed; otherwise the pipes remained empty and couldn't freeze. Besides, there was a spring nearby from which he could have carried water in a pinch.

On many a winter day, seeing the plume of smoke trailing from the chimney of his shack gave me a warm feeling as I took my morning walk. If Amos happened to be in his yard chopping wood, he always gave me the news of wood and field. The turkey flock from Flat Rock Hollow now numbered twenty-two birds, he had counted them as they crossed the old gravel road. That orphaned fawn had been seen again, grazing on the abandoned Davis farm; it showed signs of growing into a likely looking buck. Mr. Owen's son-in-law was sitting up nights with a shotgun to catch the animal, possibly a fox, that was raiding his chicken coop and evading all traps.

And on Christmas morning, of course, Amos would wave and call out, "You got those beans on a-cooking yet?" Because keeping to that ritual was what brought Amos his good luck—or so he claimed. Now others might not think Amos was so lucky, living on a level of bare subsistence like an Indian or trapper of old, but he never envied the more prosperous men of his generation who figured up their stock dividends as they sat in wheelchairs or depended on pacemakers or injections. Amos boasted of never needing to visit a doctor's office in his life—and he lived vigorously to an age of almost ninety years.

Spry and strong muscled to the last, he spent his sunset years in his own way, free and independent, working hard, roaming the woods with his rabbit dog, or boating on quiet lakes, intent always on reading directions on nature's cryptic signposts.

This winter there will be no plume of smoke drifting up from his chimney. Nobody will call out the news of nature to me on mornings when the snow crunches under my booted feet and chickadees twitter in the cedars. But come Christmas Day I'll have a pot of beans simmering on the stove for luck, and in remembrance of old Amos.

Beans for Christmas. Surely this custom brings us a message from the sturdy Ozark settlers of bygone days. It seems to say, "We have seen hard times. But with work and faith, and the help of a good solid meal of beans, we came through."